Private Forestry Policy in Western Europe

Front cover photograph courtesy of Timber Growers United Kingdom (TGUK).

PRIVATE FORESTRY POLICY IN WESTERN EUROPE

A.J. Grayson

former Director of Research,
Forestry Commission,

on behalf of the

Scottish Forestry Trust

CAB INTERNATIONAL

CAB INTERNATIONAL
Wallingford
Oxon OX10 8DE
UK

Tel: Wallingford (0491) 832111
Telex: 847964 (COMAGG G)
Telecom Gold/Dialcom: 84: CAU001
Fax: (0491) 833508

A catalogue entry for this book is available from the British Library.

ISBN 0 85198 843 1

Typeset by Colset Pte Ltd, Singapore
Printed and bound in the UK
by Short Run Press Ltd, Exeter

CONTENTS

PREFACE

Since private ownership is the main form of tenure of woodlands in Britain and the rest of western Europe, the subject of private forestry policy in Europe is important to foresters. While change on the forestry front has been great in Britain in the past decade, as it has in many other countries, and promises to continue, these shifts do not provide the only reason for commenting on current developments: the more important justification is to extend the debate on the foundations of forest policy with special reference to private woodlands and the means that governments adopt in achieving the goals of policy.

The origin of this study lies in a proposal from the Royal Scottish Forestry Society that, with recent changes in the financial arrangements influencing forestry in this country, foreign interest in investing in afforestation in Britain was likely to increase and it would be valuable to set out for comparison the incentives available to owners and investors in forestry in those countries. It was felt that foreign interest might be enhanced by the greater degree of openness flowing from moves to make the European Community more truly a single market by the end of 1992.

I am grateful to the Scottish Forestry Trust for taking up the suggestion for a study in the field of private forestry policy and for sponsoring it. The subject is one of absorbing interest because it touches on so many aspects of human behaviour, of history and of politics. This study extends the discussion beyond the boundaries of the original question. Despite the coverage that has been attempted, it has become clear in the course of the work that the whole subject of forest policy development deserves fuller enquiry than has been possible in the present review.

British forestry literature shows a marked imbalance towards biological and technical matters; the policy and economics aspects of the profession's concerns are left largely untouched by published papers. This is partly because the numbers of those with the time and inclination to carry out such research are small. But beyond this, those employed in the public service are hardly encouraged to write about, still less to criticize, public policy. The barriers are at last breaking down in this field, although there are unfortunately few examples of foresters or administrators exercising this new-found freedom. Foresters outside the forest service tend not to use the written word to convey their thoughts, while commentators from outside the profession do, often to the cost of those within, as well as the truth. It is my hope that there will be more discussion, reporting and debate on this most influential of all topics, the role of the government in setting the aims and goals for foresters and the means for them to implement policy, and that the present, slight work will act as a stimulus to others to debate, correct and amplify the views I express.

I acknowledge with thanks the help and support of Mr John Eadie, Director of the Scottish Forestry Trust, of former colleagues within the Forestry Commission and especially of a former chief, Dr F.C. Hummel, of friends in the private forestry sector working as owners, managers and agents, of forestry friends overseas who have been courteous and ever helpful in providing material and guidance, and in particular:

In Australia – Dr G. Malajczuk of the Department of Conservation and Land Management, Western Australia;
In Belgium – Professor N. Lust of the Forestry Department, University of Ghent and Mr M. Terlinden of the Royal Forestry Society of Belgium who reviewed the whole of Chapter 7; Mr P. Engels of the Royal Forestry Society; Mr R. Vanhaeren and Mr G. van de Maele of the Flemish Forestry Administration; Mr Y. Bertholet of the Walloon Forestry Administration; Mr J. Renault of the Belgian Ministry of Agriculture;
In the Czech and Slovak Federal Republic – Mr P. Rybničk of the Czech Ministry of Agriculture; Dr R. Svitok of the Slovak Ministry of Forest and Water Management;
In Denmark – Professor Finn Helles who reviewed the whole of Chapter 12, and Mr P. Holten-Andersen of the Forestry Unit, Royal Veterinary and Agricultural University; Dr N. Elers Koch of the Danish Forest and Landscape Research Institute; Mr J. Thomsen of the Danish Forestry Society; Mr K. Raae of the Danish Forest Owners Cooperatives; Mr K. Jørgensen of the Ministry of Agriculture, Messrs E. Andersen and U. Lorenzen of the Forest and Nature Agency;
In France – M. J. Guillard, former Inspector General of the French Forest Service and Professor Buttoud of the Forest Economics Branch of the National Institute for Agricultural Research, Nancy who reviewed the whole of Chapter 7; M. J. Pardé of the Centre for Forestry Research, Champenoux; M. P. Bazire, former Inspector General, Forest Service; M. P. Normandin of the Forest Economics Branch, Nancy; Messrs B. Chevalier, O. Martin de Lagarde, D. Pelissie, Mesdames M. Tregouet and I. Latournarie of the Forest Service; Professor J.L. Martres of the

Association of Timber Growers of the South West; M and Mme de Reure of the National Federation of Forest Owners; M. M. Neveux of the National Association of Regional Centres for Private Forestry; M. D. Duyck of the Normandy Regional Centre for Private Forestry;
In Germany - Mr M. Schwoerer-Böhning of the Ministry of Food, Agriculture and Forests, Bonn and Professor H. Brandl of the Forestry Research Institute, Freiburg who reviewed the whole of Chapter 10; Dr Schwenke also of the Federal Ministry; Messrs K. Giesen and W. Bachofer of the German Forest Owners' Association; Professors E. Brünig and C. Thoroe of the Federal Forest Research Institute, Reinbek; Drs G. Kent, G. Mahler and O. Schwarz of the Forestry Research Institute, Freiburg; Professor K.-R. Volz of the Faculty of Forestry, Munich University; Mr G. Pagenstert of the Rhineland Agriculture Board;
In Hungary - Mr F. Gerely of the Ministry of Agriculture, Budapest;
In Ireland - Mr B. Hussey of Woodland Investments Ltd and Mr J. Gillespie of the Forest Service who reviewed the whole of Chapter 6; Dr N. OCarroll, Mr F. Mulloy and Mr E. O'Flannagåin of the Forest Service; Mr M. Bulfin of Teagasc;
In the Netherlands - Mr R. Busink of the Forest and Nature Service, The Hague, Mr E.J.G. Swellengrebel, formerly of the Forest and Nature Service, Messrs R. van Woudenberg and A.J.J. Bakker of The Netherlands Association of Forest Owners and Mr B. van Vloten, woodland owner, all of whom reviewed the whole of Chapter 9; Mr A. de Wit of the Centre for Agrobiological Research, Wageningen; Mr P. Hinssen, De Dorschkamp Institute for Forestry and Urban Ecology, Wageningen;
In New Zealand - Mr J. Valentine of the Ministry of Forestry; Hon. K. Shirley of the NZ Forest Owners' Association; Mr J. MacBride of the NZ Farm Forestry Association;
In Norway - Mr O. Aalde, Forest Service, Mr E. Bjorå of the Norwegian Forest Owners' Association and Professor J. Eid of the Forestry Department, Agricultural University of Norway who reviewed the whole of Chapter 13; Mr L. Askheim, Forest Owners' Association; Messrs I. Ekanger, T. Opheim and A. Scheistrøen of the Forest Service; Professors A. Svendsrud, S. Nersten and H. Frivold and Messrs S. Baardsen and H.F. Hoen of the Forestry Department and Professor A. Lunnan of the Agriculture Department, Agricultural University;
In Poland - Dr S. Zajac of the Forest Research Institute, Warsaw;
In Russia - Acad. N. Moiseev of the All-Russia Research Institute for Silviculture and Mechanization of Forestry;
In Sweden - Professor L. Lönnstedt of SIMS, Swedish University of Agricultural Sciences and Mr C.-G. Dahlin of Skogsstryrelsen, Jönköping who reviewed the whole of Chapter 14; Mr Holm, Jönköping, Mr U. Didrik of Skogsvårdsstyrelsen Uppsala; Mr U. Österblom and Mr S. Hogfors of the National Federation of Forest Owners' Associations, Stockholm; Professor

L. Hultkranz, Department of Economics, University of Umeå;
In Switzerland - Mr R. Badan, Head of the Forest and Vineyard Service of the city of Lausanne who reviewed the whole of Chapter 11; Mr U. Amstutz and Mr E. Affolter of the Swiss Forestry Association; Mr D. Schmid of the Federal Forest Service, Bern; Professor F. Schmithüsen, Drs M. Sieber and T. Fillbrandt of the Forestry Department of the Federal Institute of Technology, Zürich; Dr P. Schmid–Haas of the Federal Institute for Forest, Snow and Landscape Research, Birmensdorf; Mrs C. Gussmann of the Tax Information Office, Bern;
In the United States - Dr F. Kaiser of the USDA Forest Service, Washington, DC; Dr F. Cubbage of the SE Forest Experiment Station, Forest Service; Professor P.V. Ellefson of the Department of Forest Resources, University of Minnesota; Professor P.W. Adam, Department of Forest Engineering, Oregon State University;
In the European Commission - Messrs J. Wall and K. Jakobsen of DG VI, Mr R. Flies of DGXII, Brussels.
In the Economic Commission for Europe - Messrs T. Peck and P. Schröder;
In the Comité Central de la Propriété Forestière - M.P. Gathy and Baron A. de Jamblinne de Meux;
In the European Confederation of Agriculture - Mme M. Ré;
In the Confederation of the European Paper Industry - Mr E.D. Franke.

None of the above named bears responsibility for errors of fact or interpretation which remain those of the author alone.

Finally, I have to thank my daughter, Cecilia, for German translation, and my wife, Jessica, for her support and forbearance during the preparation of this study.

A.J. Grayson
Farnham, England
October 1992

1

INTRODUCTION

The Current Background to Private Forestry in Britain

The late 1980s and early 1990s have proved a particularly testing time for private woodland owners in Britain, as elsewhere in Europe, for a number of reasons, partly economic, notably recent high prices of capital and historically low wood prices, and partly because of recent or impending events affecting the policy environment. The latter involve changes in the aims of government forestry policy towards private woodland owners in many countries and in some instances in the means by which policy goals may be achieved. The following are among the most important of recent or ongoing events affecting British woodland owners in particular:

1. The removal of forestry from the income and corporation tax regime in 1988 combined with the substitution of higher levels of grant which has changed the attractiveness of investment especially in afforestation, while new classes of potential investors have yet to emerge.
2. Depressed timber prices, 1991–1992 levels of standing tree prices in real terms probably being as low as at any time since the 1930s depression.
3. The continuing shift towards a demand for the greater provision of non-market benefits, notably in fields such as nature conservation and recreation.
4. Emergence of new concepts in the British scene such as community forests.
5. Changes in agricultural support in the light of the reshaping of the Common Agricultural Policy, the negotiations under the General Agreement on Tariffs and Trade and the combined effects on land price and public attitudes on the use of agricultural land.
6. The advent of the unified market of the European Community and its effects on product (roundwood, sawnwood and wood panels) and to a lesser extent on input (for example professional staff) markets.

7. The uncertainties associated with the impact on economic conditions generally of developments in European Community policy.
8. Progressive sale of the Forestry Commission's forest holdings.
9. Wider political concern with the global significance of forests, including moves to give greater protection to existing forest in all parts of the world and the likely effects on timber supply that this will have, and the use of forests and wood products as carbon stores.

Some of these developments are peculiar to Britain, some are engendered through membership of the European Community, and others arise through world events and concerns. Further issues may emerge in other countries and these may come to influence attitudes in Britain. One arm of the present study is concerned with the identification of policy issues in forestry and in particular with the weighting accorded to different objectives in countries.

The other subject of attention is implementation. A wide range of methods is deployed in support of government policies bearing on the private forestry sector, ranging from legal regulation through financial incentives such as grants to the provision of free advice. The balance, costs and effectiveness of the various means of support are obviously of interest if the best use is to be made of public resources devoted to the promotion of forest policy.

The Relevance of European Experience

European experience is important because of the broadly similar cultural backgrounds, political philosophies and comparable levels of real earnings and patterns of demand of continental countries to those of Britain. But within Europe there is a diversity of traditions and forest histories, and new trends and different emphases (for example the growing concern to create 'near to nature' forests) are emerging among the aims of policy. In practice, differences in the sets of policy objectives adopted in countries may not be so large. On the other hand, it is reasonable to expect there to be rather greater variation in the range and weights given to each of the tools used by government to implement policy towards private forestry.

Visits have been made to nine countries in connection with the study and the help of correspondents has been enlisted in the cases of seven others. Population, land and forest area statistics for the countries referred to are presented in Appendix I.

Purposes of Study

It is impossible to do justice to the long and complex history of forestry in the selected countries and to the forces that shaped one or other aspect of policy and are still shaping them. Accordingly, the aim is to provide some very general background and to condense statements of the policies in force. This study does not consider the question of the desirability of private forest ownership but accepts it as given.

In addition to the actual content of policy, the methods of formulating policy differ among countries. One feature of the formulation process is that the range of centres of influence is growing wider. This aspect is indeed so important that it may be regarded as the largest single new development in the whole policy scene from conception of ideas to practical action on the ground to have emerged in recent decades. This situation is well enough known in Britain; whether the appropriate methods of response are used may be questioned. It has not proved possible in the time available to give special attention to this topic which requires wide consultation if any useful findings are to emerge: nevertheless some proposals are set out for consideration in Chapter 18.

In relation to implementation, it is has been considered unnecessary to detail the regulatory powers of the forest authority and other departments of government. Instead attention is concentrated on details of financial incentives provided.

In sum, the purposes of the study are as follows:

1. To describe the main strands of policy towards private forestry in a number of countries that appear relevant to the British situation.
2. To identify methods of policy formulation and trends in the content and balance of objectives of private forestry policy that may have implications for the development of policy in Britain.
3. To describe methods of implementing policy and, as far as possible, quantify public support of private forestry.
4. To indicate any changes in implementation methods that appear appropriate in Britain.

2

THE DEVELOPMENT OF POLICY

THE NEED FOR A FOREST POLICY

An essential point to make about policy is that it is not necessary to have a policy if the industry concerned is behaving to the satisfaction both of those engaged in the industry and of society. Governments do not generally produce policies for the establishment and management of sweetshops: in this case the market satisfies society's needs quite competently. Forestry, it is argued, is one area where the market is littered with failures (HM Treasury, 1972; Price, 1989; Simula, 1990; Pearce, 1991). Assuming this to be so, the situation provides both a necessary and a sufficient reason for the creation of a policy.

Westoby (1989) has strongly urged the case for a declared national forest policy for a further reason, namely as a means of ensuring that forestry activity will continue to be given political support.

DEFINITION OF 'POLICY'

Clawson (1979) defined policy as 'all public actions which significantly affect the use and management of forests'. Such a statement, while true, is, however, capable of applying to a policy which in whole or in part may not need to exist. Here it will be assumed that policy has a definite purpose and, following the example of the wording of the European Community's policy on industrial competition, this may be stated thus:

> policy aims to facilitate and encourage practices which have as their effect the creation and maintenance of such forests as satisfy national wants.

Content of Policy Statements

Analysis of the Geneva Declaration of 1985 on (government) forest policy (Hummel and Hilmi, 1989) shows that this brings together three policy concerns. These are the forest policy of the state without regard to the class of owner concerned, the policy towards one particular class, for example that of the individual private owner, and the various implementation techniques. In view of the way in which the word 'policy' has to do duty for so many concepts, it is useful to discuss in more detail the meaning to be attached to the term (the qualification 'government' is tacit in what follows).

At one level the objective of all policy is the same in a democracy; it defines the aim as the greatest good of the greatest number, and although this statement is a contradiction in terms, the cliché continues to attract. The aim may be voiced in rather more operationally significant terms, such as balancing the demands of society with the capacity of the forest resource (Kanowski, 1991, quoting van Maaren), although in this formulation the idea of extending the forest as one means of altering capacity is not immediately apparent. Again, it may be expressed at greater length but with more precision as in the section headed 'General principles' in the statement developed by Hummel and Hilmi (1989, p. 116). This valuable statement sets out measures needed to be taken to achieve aims in the various fields of wood production, nature conservation and environmental protection, recreation and forest industry. It also links with each heading specific ideas as to implementation, for example who should pay for the works necessary to meet a particular requirement.

The Statement goes on to say that 'Forestry policy should be concerned not only with the identification of objectives but also with the means of implementation and updating.' It may be argued that properly the matter of implementation is extraneous to policy. This is however an unrealistic view. Government, managers, special interest groups and the general public share concerns about the methods used to implement policy along with their concepts of appropriate objectives.

It is desirable that policy should distinguish between government policy on forestry generally and policy towards particular groups of owners. It is usually only practicable to make broad statements of intent in policy statements, although exceptionally a very precisely stated aim may be appropriate, as in many countries which specify *inter alia* that the area of forest in a given region shall not decrease. To proceed further usually requires that the particular ownership category be identified, and the means of implementation then related to that ownership class. In theory, the position is entirely symmetrical between the state forestry enterprise and private owners. In practice, the general statement of policy often does duty for the state enterprise, the means of policy implementation for which tends to be taken for granted. There are two principal reasons for suggesting that policy towards specific groups of owners should be distinguished. The first is that it is pointless to establish objectives whose implementation is impracticable. Thus it may become apparent that only one category of owner such as the State itself is in a position to effect certain aims, or to effect them at economic cost in terms either of resources or of public expenditure – the inducements necessary to elicit this response from local authorities or from private owners who would otherwise have to be paid being exorbitant. Examples might be objectives for private woodland owners such as the production of veneer oak, the establishment of community forests or a particular scale of new planting. Second, a policy consisting only of a statement of objectives for a given ownership group does not constitute a policy if some form of government intervention is necessary in order that objectives be achieved. The specification of the implementation measures proposed is then an essential part of the policy.

The conclusion that it is necessary to associate a statement of the means with the statement of objectives has the great merit of emphasizing the desirability on economic grounds of linking the two aspects. It is clear that too rarely have the costs of policies, towards either state or private forests, been compared with their effects. This is the principal theme of the field of enquiry called 'policy analysis' which is referred to in Chapter 3 and discussed more fully in Chapter 17 and, with particular reference to extension services and the decision to cut timber, in Appendix III.

Can a Forest Policy Succeed?

It is a striking feature of forestry that those on whom policy bears can almost never achieve what is desired. The nature of forestry is such that the fruits of past policies only become available long after their goals may have been superseded, or their objectives altered by new events. This aspect puts a special onus on the two agents involved. First, administrators and politicians must be educated to appreciate that there are very long lags between the implementation of actions designed to produce certain results and the actual delivery of those results. Second, owners are wise to recognize that voters and their political representatives are liable to change the positions of policy goalposts frequently relative to the practicable rate of change of forest design. Owners may well adopt an attitude of caution instead of moving quickly to new targets and techniques.

The Prediction of Issues

The term 'issues' is used to denote subjects considered relevant by participants in the policy-making process: such issues may concern objectives or methods of implementation. Policy making would be an easier, as well as more successful, process if it were possible to predict the issues that are likely to be influential some years, or in the case of forestry, several decades, ahead. There might then be a better chance of making timely adjustments in programmes of support and in the plans of forest owners, for the more abrupt changes are, the more costly they are apt to be. Thus if there had been a series of leading indicators of changing attitudes to species choice in afforestation in Britain, there might have been more timely preparation for the surge of interest in broadleaves, especially native species, which appeared in the early 1980s. Or the differences over a particular group of peatlands that so dramatically disturbed the official conservation agency as well as afforestation activity in the north of Scotland might have been averted. The prediction of trends in the future economic environment in which foresters are likely to have to work shows that long-term predictions can be useful guides to action. But efforts to predict changes in aims are less frequently made. Major swings spring fully armed rather than emerge in some clearly coherent fashion. In contrast to macroeconomic forecasting, objective data for well-defined variables are much more difficult to identify in the field of policy and relationships remain undetected.

But simply to pose the question of emerging issues indicates the difficulty of the task. The process of detecting changes in what a nation or a particularly influential group requires from some group such as the body of private forest owners is unlikely to be achieved by the normal processes of scholarship since it is in the nature of these matters that the relevant data are not readily available. Indeed, the fact that historians can argue many years after the event which were the relevant factors leading to a particular incident, gives cause for doubt about man's ability to identify underlying trends in time to have any operational use. Even an adequate basis of classification of events is missing. A desirable aim would be to see if it is practicable to detect turning points in factors determining policy. Though ingenious ideas have been paraded in the pages of journals such as *Futures*, including ones concerning policy in relation to aspects of the physical environment, tests are by their nature hard to devise. Some novel work relevant to forestry has been carried out by Williams (1991) who attempts to use quantitative measures of policy variables in order to indicate possible economic changes following economic reform in the former Soviet Union. Short of such an approach, the most sensible course appears to be to observe what is happening elsewhere, not only in forestry, and to attempt to report developments which may be repeated, or at least influence thinking, in Britain. While this approach is a comparatively low-risk one, the lead time involved is bound to be rather short.

Steps in Policy Development

There are relatively few works devoted to the analysis of forest policy and still fewer to the steps in its initial or periodic reformulation. Ellefson (1992) describes the administrative steps in policy development. These have as their point of departure agenda setting, which leads to the government agenda which in turn is succeeded by the formulation of options. As Ellefson and Lyons (1989) observe in their valuable review of policy development and programme administration, very little attention has been devoted to agenda-setting as a strategic process in forestry. Agenda-setting processes (Schmithüsen and de Montalambert, 1991) may stem from the forest sector itself, from new or changing interests of groups outside forestry or from effects of other government policies. Even where there is agreement on the agenda there is still no means of objectively ranking items in terms of importance or analysing how they came to be regarded as important. Foresters have not been active in this area, although some students of public and company administration claim to have insights into the subject. On the forestry front, the existence of at least one (American) journal devoted to the study of history in relation to forestry, *Forestry and Conservation History*, and discussions in other parts of the world by Dargavel (1991) are welcome indications that these studies are gaining in interest.

It is not the purpose of this study to provide a review of the various steps in policy formulation. For this the reader is referred to Hummel and van Maaren (1984) and, for overviews with a more distinctly American flavour, Cubbage *et al.* (1993), Ellefson (1992) and Worrell (1970). However, recent decades have seen the emergence of two elements in policy discussions which are so influential as to demand attention. One concerns the burgeoning of pressure groups as a form of public participation, the other the government response.

The Activity of Pressure Groups

Recent decades have seen a dramatic growth in the activity and influence of pressure groups concerned with countryside matters. The influence they command may be very large relative to the number of active participants. The impact concerned has come from a direction which can be regarded as quite unusual by the standards adopted by most people engaged in economic activity: not that of the people engaged directly, the stakeholders in business management parlance, but of one or more sets of outsiders. Indeed these outside groups, or more specifically single interest groups, have in many cases come to be regarded as stakeholders, not because of any conventional legal claims as interested parties but because their interest has grown to have political weight. In Britain an early illustration (1936) of the influence of such outside bodies, or lobbies, was the criticism by the Ramblers Association of Forestry Commission afforestation in areas of high amenity interest, notably parts of the central Lake District. Of course, it did help the case to be able to pray in aid the Archbishop of York in whose archepiscopate the origin of the *cause celèbre* lay.

In his discussion of the 'New Pluralisms', Drucker (1989) draws attention to the differences in the way that society and what he terms the 'polity' have come to be influenced, even dominated, in modern times compared with a century ago. He claims that each area of enterprise is now marked by the existence of power centres outside and separate from government. In the case of the polity, Drucker draws attention to the pluralism of single-cause groups which, though small minorities, are well disciplined and capable through leverage of obtaining results otherwise achieved only by numbers or 'rational democratic persuasion'.

The use of leverage echoes the organization of the Bund der Landwirte in late 19th century Germany to achieve its aim of reinforcing agricultural protection. Tracy (1989) refers to this creature of the Prussian landlords – the Junkers – as a new type of pressure group. It organized demonstrations, distributed propaganda, issued manifestos and ordered its members to boycott traders who did not give support. Its principal means of action, however, was political. In elections the Bund gave support to candidates willing to suscribe to its programme. It kept a strong hold over all members of Parliament elected with its support; anyone who departed from its line usually found his political career at an end.

The movements to which Drucker refers are extra-parliamentary and their importance, he argues, is that they increasingly paralyse and tyrannize political life in developed countries. The issues they support are diverse. Among the examples quoted of successful lobbies from earlier periods are Pulitzer and Hearst in 1898 leading the United States into the Mexican War, and, *par excellence*, Lenin. Prohibitionists provide another dramatic instance of the power of an organized selection of political candidates. Nowadays 'social' as opposed to overtly economic movements dominate, examples being objections to abortion, rice growing in Japan, and support of single issue 'green' policies. Drucker is silent on whether the situation can be controlled, though it should be noted that a wholly extra-parliamentary activity, namely secondary picketing, was banned in Britain in 1985 and with it the influence curbed of those regarded as too distant to be considered stakeholders in the particular industry affected.

There is an additional element that may be considered essential to the success of today's pressure groups. Lowe (1987) argues in relation to environmental matters that it is no longer the leaders of organizations alone but much more crucially the media that convey, and, he might have added, dramatize, the message of a particular group to the public at large.

Pressure groups relevant to the formulation of policy today appear in many fields that affect or may affect forest owners and through them the practice of forestry. The study of the nature and effectiveness of such groups in forestry is primitive. Metz (1986) provides one illustration of the development in western countries of what she terms 'neocorporatist' systems. These rely not as previously on competition between interested groups, but on the association of subordinate groups with the political process. Metz draws attention to the mutuality of interest of state, forest industry, forest owners and trade unions in the construction of Finland's 'Forest 2000' programme in the early 1980s. The history is an intriguing one, not least for the reasons that the original shared concern, namely a scarcity of wood for industry, has now evaporated, and the changing character of forest owners, who are increasingly town dwellers, will in itself demand the emergence of a new struggle for concensus.

The Response

The development of government policy, often leading to legislation, in a particular field such as civil rights, education, labour or other theme owes its existence to the efforts of past generations of campaigners. The main objection to single-interest groups as at present constituted is not necessarily their success, but the way in which their case and information about the issue are presented to their audiences. Westoby (1987) in a chapter headed 'Forestry, foresters and society' suggests that in large measure foresters have only themselves to blame for not informing politicians and administrators adequately of the facts about forestry and its potential contribution to society. Hummel and van Maaren (1984) emphasize the need for the supply of adequate information and wide consultation with the general public in addition to opinion formers and officials.

Forestry still suffers from a partial, that is, biassed as well as incomplete, revelation of its objectives, goals and achievements. It is essential that the provision of information is not subjugated to the desire to present reality in the best light. The need is for information which is educational rather than of a public relations kind. The subject of consultation as one feature of the government's responsibility in developing forest policy is returned to in Chapters 5, 16 and 18.

Other Considerations in Policy Development

The allocation of priorities among objectives

Many criteria are available by which to choose the weighting to be attached to the different outputs of a given policy. A continual struggle is waged between protagonists of technical and ecological criteria and those favouring economic and social measures. Much of this debate is redolent of the barren arguments which were once a feature of the choice of management regime in a given forest. The hope is that a balanced economic set, incorporating equity where needed as well as efficiency measures, should be adopted. The expectation is that this will remain difficult to achieve over the whole field, especially in relation to wildlife conservation. The desirability of establishing a single measuring rod, money, for the efficiency criterion is undoubted.

Integration with other policies

Forestry policy should clearly be compatible with other policies. It would be a counsel of perfection to suggest that they must be integrated initially since not all possible fields of conflict can be foreseen, nor may be it be politically apt to raise potential sources of conflict. It is at the stage of implementation that strains become obvious, and in this connection there is an obvious merit in regional planning both for management of existing land uses and of changes in land use (Goodstadt, 1992).

Incrementalism

A common approach to policy making or policy implementation consists in tacking on one additional arrangement after another to an existing policy or scheme. Examples are the adjustment of the terms of the French Fonds Forestier National and the progressive manipulation of the British Dedication Scheme between 1947 and 1981. The alternative of radical change is regarded with distrust by many foresters, as evidenced by the concern expressed in Britain following the breakdown of bipartisan accord on the roles of state and private forestry in 1980 when the law allowing the sale of state forests was passed. But incrementalism may imply that policy instruments get out of hand and no longer serve the intention of current policy adequately. Studies on this topic are very rare, partly because of the absence of controlled situations, either in space or time, which might allow identification of the effects of a given increment.

References

Clawson, M. (1979) *The Economics of U.S. Nonindustrial Private Forests*. Resources for the Future, Washington, DC.

Cubbage, F.W., O'Laughlin, J. and Bullock, C.S. (1993) *Forest Resource Policy*. John Wiley and Sons, Chichester.

Dargavel, J. (1991) Record of meeting on history of the forest economy of the Pacific Basin. *IUFRO News* 20, 16.
Drucker, P.F. (1989) *The New Realities*. Heinemann Professional, London.
Ellefson, P.V. (1992) *Forest Resources Policy*. McGraw-Hill, New York.
Ellefson, P.V. and Lyons, J.R. (1989) Policy development and program administration. In: Ellefson, P.V. (ed.), *Forest Resource Economics and Policy Research*. Westview Press, Boulder, pp. 228–240.
Goodstadt, V. (1992) Constructing an indicative strategy: the Scottish experience. Paper to conference on 'New Forests for the 21st Century'. Forestry Commission, Edinburgh.
HM Treasury (1972) *Forestry in Britain: an Interdepartmental Cost/Benefit Study*. HMSO, London.
Hummel, F.C. and Hilmi, H. (1989) Forestry policies in Europe. *FAO Forestry Paper 92*. FAO, Rome.
Hummel, F.C. and van Maaren, A. (1984) Policy formation. In: Hummel, E.C. (ed.), *Forest Policy, a Contribution to Resource Development*. Martinus Nijhoff, The Hague, pp. 283–303.
Kanowski, P. (1991) Review of 'UK forest policy into the 1990s'. *Forestry* 64, 83–84.
Lowe, P. (1987) Environmental concern and conservation politics. In: Whitby, M. and Ollerenshaw, J. (eds), *Land Use and the European Environment*. Belhaven Press, London.
Metz, A.-M. (1986) Influence of owners as an interest group in achieving the forest policy goals in Finland: the programme 'Forest 2000'. In: Tikkanen, I. (ed.), Analysis and evaluation of public forest policies. *Silva Fennica* 20, 286–292.
Pearce, D. (1991) Assessing the returns to the economy and to society from investments in forestry. Paper No 14, *Forestry expansion: a study of the technical, economic and ecological, factors*. Forestry Commission, Edinburgh.
Price, C. (1989) *The Theory and Application of Forest Economics*. Basil Blackwell, Oxford.
Schmithüsen, F. and de Montalambert, M.R. (1991) Current trends in forest policies. Position paper, *Proceedings of the 10th World Forestry Congress*, Paris.
Simula, M. (1990) Structural change of the forest sector in a global perspective – challenges for forest policy design. *Proceedings IUFRO Division* 4, 450–455.
Tracy, M. (1989) *Government and Agriculture in Western Europe 1880–1988*. 3rd edn. Harvester-Wheatsheaf, Hemel Hempstead.
Westoby, J. (1987) *The Purpose of Forests*. Basil Blackwell, Oxford.
Westoby, J. (1989) *Introduction to World Forestry: People and Their Trees*. Basil Blackwell, Oxford.
Williams, C.M. (1991) Cross impact analysis of economic reform and political reform in the Soviet forest industry. Typescript. Forest Economics and Policy Analysis Research Unit, University of British Columbia.
Worrell, A.C. (1970) *Principles of Forest Policy*. McGraw-Hill, New York.

3

THE IMPLEMENTATION OF POLICY

INTRODUCTION

The foregoing discussion of issues and influences on policy formulation indicates the poverty of our understanding of the reasons both for the emergence of certain topics as policy issues and for the processes of policy formulation that lead to some issues becoming incorporated into policy objectives. The position is quite different when the means used to implement policy are considered. In human history generally, much more is known about the minutiae of campaigns and more is usually understood about tactical affairs than about the reasons for the strategy adopted. So it is in studies of economic and social policy, and in particular of forest policy.

Interactions with policies in other fields

Forest policy, and particularly its implementation, can never be considered on its own since forestry is carried on in the context of a country's or region's current economic and social climate which itself is heavily influenced by government economic, including fiscal, policy, by legislation on property and control of land use. In recognition of this, some studies have attempted to comprehend the links between forest policy and other aspects of government policy (Riihinen, 1986). However, even in a country like Finland with such a large proportion of its population engaged in forestry and forest industry it is salutary to remember the caution expressed by Tikkanen and Vehkamäki (1990) who, noting the desirability of combining parallel measures in fiscal, agricultural and forest policies, recognize that the prospects of such coordination are low since 'forestry has a rather low priority in Finnish political decision making'.

The Target for Policy: the Private Forest Owner

Literature review

Census statistics for forestry rarely identify the owners of the forests surveyed except in the broadest terms (state, local authority, private). Special surveys are mounted to describe owners: an example of a comprehensive statistical enquiry is that by Ministère de l'Agriculture (1987) on private woodlands in France. Among the studies that aim to detect and measure reactions to public policy incentives such as grants, tax concessions and extension services is the group that attempt to discover in what ways owners differ. If successful, these should provide an indication of points at which government intervention may be expected to be most effective.

A first approach is a subjective descriptive one. Thus, Byron and Boutland (1987), reviewing policy towards private owners in Australia, asserted that it would be 'naive to expect one package to suit all categories of owner'. They quickly separated industry from farmers and syndicates. In the following discussion, attention is concentrated on the owners of what is termed 'non-industrial private forest' or NIPF in the American literature and, by adoption, in Scandinavian writing also. Here the initials PF are used as sufficiently distinctive.

In France, Normandin (1987) used a purely objective statistical technique (cluster analysis) to distinguish private owners by Département, this being the lowest level at which relevant data were available from the Ministry of Agriculture's statistical report ESSES (l'enquête statistique sur les structures économiques de la sylviculture, Ministère de l'Agriculture, 1987) which characterized the numbers of private owners concerned and their financial position. He also made a characterization using woodland census data by reference to the stands they owned, operations conducted (including timber sales) and the intensity of management. Finally, a combined classification was produced covering both ownership and management features. The use of this work has not been described but clearly it could, if desired, form the basis of regionally differentiated systems of policy intervention.

Kurtz and Lewis (1981) adopted a revealing approach by using a psychological testing technique. This relied on the degree of agreement individual owners showed with alternative statements about their motivations (money return, investment aims, etc.) in relation to timber production, wildlife, grazing and preservation. As a result four sorts of owner were identified, ranging from 'timber agriculturalist', in effect timber farmer, to 'forest environmentalist'. The writers suggested that PF owners in the different categories should be treated in different ways so far as advice on timber growing was concerned. Young and Reichenbach (1987) employed a social/psychological model to discern the root cause behind owners' behaviour, namely their intentions. They developed a classification clearly separating those who intended to cut their woodlands at some time from those disinclined so to do. The relevance of the distinction is that traditional forest incentive programmes, such as grant schemes, are pointless so far as 'non-intenders' are concerned, since they do not believe it desirable to supplement their income by cutting wood. Similarly traditional education and extension programmes are unsatisfactory in relation to this class of owner. If it is government's purpose to mobilize more wood, the need is to change the beliefs of 'non-intenders' by, for example, increasing the association in an owner's mind between timber cutting and the satisfaction of domestic firewood needs, or between timber cutting and diversifying the habitat for wildlife. Of course applications of the findings may be in a totally different direction if the purpose of government is to persuade PF owners to act in favour of conservation and the minimization of wood output.

Most of the studies on the classification of owners have, however, been based on more direct approaches to actual observed behaviour, and that mainly in connection with either silvicultural or logging operations. This concentration reflects governments' paramount concern relative to woodland owners with the means to be used of increasing the amount of wood-growing activity and raising the efficiency of wood production. An interview survey aimed at discovering different investors' attitudes to the current Forestry Commission planting grant scheme was conducted by Firn Crichton Roberts (1990). The five categories of investor included 'latent investor' and 'estate adviser' in order to ensure coverage of a target audience of likely investors in new planting. Similarly, Yoho (1985), in a perceptive review of private investment strategies, identified five classes of NIPF investor in forests, including custodial investors who have inherited a property but who have no interest in the management of the asset, speculators and 'true investors'. Each class was distinguished logically on grounds of different responses to government policy initiatives such as grants or tax regimes. Greene and Blatner (1986) followed the kind of objective approach used by Normandin but applied it to individual owners. By investigating the age, educational attainment, farming interests, etc. of 1400 Arkansas owners, they developed a formula for discriminating between people who were woodland managers from those who were not.

Although some of these studies must appear to be of the 'sledgehammer cracking a nut' variety they are intensely relevant to the understanding of private owners' behaviour. An intrinsic deficiency of all the studies referred to is that, as Riihinen (1986) observes in relation to policy analysis generally, too little attention is paid to dynamic effects. Usually these will be of the nature of changes in attitudes apparently independent of external economic influences. It is, however, vitally important to remember this aspect: attitudes change and early detection of such changes is desirable if rational decisions are to be made about shifts in public policy and support.

Although it is obviously convenient for analysts to consider the behaviour of a broadly homogeneous group of owners, the reality is that practical reasons of administration usually mean that government operates a single scheme taken up by people from widely different groups. It is then even more necessary to attempt to identify the reactions of the various kinds of owner to the policy instrument.

Establishing Goals

As Ellefson and Lyons (1989) indicate, the ways in which goals of policy are set is most obscure. Sometimes the considerations leading to targets may appear naive to the point of absurdity. Thus in 1973, the government announced (Forestry Commission, 1974) that the future annual rate of planting set for the Forestry Commission, i.e. State forests, would be up to 55,000 acres (*c.* 22,300 ha) for new planting and replanting combined. The rate of replanting would then be determined by fellings which are in turn broadly determined by the pattern of variation in planting rate decades before. There is clearly no logic for the assumption that the need for new planting in future should vary inversely with the new planting of five decades earlier; in fact the goal is to be interpreted as a compromise between the desire to encourage new planting and fiscal prudence. Similar comments about their arbitrary nature could be made about the rates of new planting which the British government established for the private sector more recently: the basis for the sizes proposed for the new community and English Midland forests (Forestry Commission, 1990) is unknown.

Implementation Tools

Johnson (1990) has compiled a comprehensive list of intervention tools. Some are intersectoral, such as trade and environmental regulatory arrangements, land use incentives, or public finance variation such as relief for property tax on agriculture and forestry. Others are intrasectoral, and include input subsidies, tax allowances, and forest management regulations. One of the tests of a government's resolution is the money it is willing to provide, or the tax revenue it is willing to forego, to encourage action on the ground. A main purpose of this study is to present the tax and grant arrangements applying in a number of countries as at 1992.

Tax

The purpose of this section is to introduce some main features of taxation and in particular of forestry taxation as a preliminary to the descriptions and partial evaluations of individual country systems in Chapters 5–14.

Cautionary remarks

Taxes, like pesticides, have to be handled with care. Their effects may spread far from the original target. At any one time, their form often represents the accumulation of steps the origins of which are long forgotten or no longer applicable. The existing system may quite possibly represent the result of less than well-considered judgement. Indeed, Klemperer (1989), writing of the position in the United States, went a stage further in arguing that forest tax goals were poorly defined or conflicting in purpose. Allied to this point is the general consideration that change, and especially abrupt change, is not conducive to efficient management. Rational expectations are the basis for management, however vague and elusive the values concerned may be, but changes that are unexpected, such as the introduction of a wealth tax or the removal of forestry from income tax, clearly fall outside such expectations and accordingly lead, on average and over a period, to poorer decisions.

The incidence of taxes is not necessarily consonant with the current intentions of government. In the light of his major survey of ECE countries, Hart (1979) concluded that there was a significant difference between the intended and the actual effects of taxation. And principles, once regarded as hallowed and undeniable, are no longer so regarded. The criterion of 'ability to pay', in the sense of payment out of income, as a prerequisite of any fair tax system is easy to establish in relation to income tax and is indeed followed in most countries. In Britain the personal, so-called 'poll', tax introduced as a replacement for a property tax raised the issue acutely. However, it is much more difficult to interpret and apply when account is taken of the fact that wealth confers economic power. Equity is paraded as a desideratum and elevated by some to the status of a principle, horizontal equity referring to the *equal* treatment of people of comparable taxable capacity, vertical equity referring to *appropriate* tax treatment of people with differing taxable capacity. It is noticeable, however, that for reasons as much of practicality as anything, more countries have come to rely on indirect taxation in place of income tax although its impact is regressive, that is its incidence is relatively higher on those with low incomes. Another concept is that of neutrality, which was popularized by the work of Meade and the Institute of Fiscal Studies (Kay, 1989) and which by 1988 was accepted by the Chancellor of the Exchequer who saw that his duty was 'to see that, as a general rule, peoples choices were distorted as little as possible through the tax system'. It is interesting to apply the test to forestry. There may be sound reasons for non-neutral treatment of forests, for example because the market does not provide a measure of certain benefits, but an outstanding issue clearly arises since unless these benefits are generated automatically, other regulatory devices have to be imported to ensure the delivery of benefits. Secondly, it is interesting to note that under the current system of income taxation in Britain, owners of revenue-producing woodland established many years before are not subject to tax in the way that owners of other businesses are. It is clear that so-called 'principles' of taxation are often temporary totems rather than dependable supports.

The role of taxation and the relation with forestry

Taxes have three functions: (i) to raise money for the taxing authority; (ii) to redistribute income and wealth, and (iii) to encourage particular forms of behaviour by either consumers or producers. The last is clearly the motive of governments which have specific forest policies in adjusting taxation of forest owners; and in so doing to introduce difficulties simply by the fact of differentiating forestry from other forms of activity.

Governments face some extreme difficulties when judging the appropriate systems of taxation of forests. For if the need for a forest policy arises, as already argued, because of market failure, then incentives to provide outputs nearer to the social optimum may be provided in two principal forms: grants (with which may be grouped materials and advice given *gratis*) and tax reliefs. If government wishes not just that the private forestry sector should prosper, but also should increase its investment, it must offer even more substantial incentives. Simultaneously the opportunities for tax evasion or legitimate avoidance increase. There may also be effects on asset prices through tax shifting (that is the impact of the tax is partly shifted from the taxpayer to another economic agent) which may well be detrimental to the achievement of policy goals. The choice and manner of applying these tools constitutes one of the most intriguing and testing fields for the application of forest economics. In such an intensely behavioural field as the maintenance of long-lived investment, and even more acutely in relation to investment in new forests, such aspects as confidence, shifts in the general philosophy underlying taxation, the changing availability of other investments, and changing cost and price conditions in the sector, all mean that the current regime of tax and direct support can never be everlasting but must adjust to the realities.

A second point to note in relation to forestry (in this connection, state forest enterprises cannot be ignored because they may be subject to corporate income tax and/or property tax and therefore have to be included) is that the value of the sector's total potential tax revenue is usually small. Accordingly it might be argued that the sector could be left to take the rough with the smooth when tax changes are introduced. But this is not the experience of any country, not excluding those where forestry contributes only a small proportion of national income or other measure such as land area. Indeed the wonder is that the forestry lobbies in so many countries with a small forestry sector have been so influential in achieving favourable changes in the regimes applied. It is of course possible to argue that since the maximum tax revenues available are small it matters little to the Exchequer whether concessions are given to forestry or not. This may well have been an element influencing the decision of the British government in 1988 to remove forestry from income tax but such occasional events apart, the general experience is that tax authorities do not apply a *de minimis* rule. Although they will listen to the blandishments of particular sectors, they are well aware of the tax evasions and economic distortions that may arise if too liberal concessions are given. A side-effect of the small scale of the sector is that, as with other statistics, data for forestry on tax revenues are often grouped with figures for other businesses, thus hampering analysis.

Unnecessary obfuscation and confusion has been caused by repeating the statement that trees are both factory and product. It is assumed that, *ipso facto*, this implies that forests require or deserve some different tax treatment from that applied to the ordinary run of commercial enterprises. Although the statement is true, it conveys no important information which marks forests or indeed individual even-aged stands as being any different from a conventional factory. Each has a market value as a going concern whereas the apparatus of production remains in being, each produces income after an interval from the time of laying down the first brick or seedling and the pattern of income over time varies over the life of each sort of 'factory'. It happens that forestry is characterized by a higher capital-output ratio than most other activities.

Lastly, it is important to recognize that words that are similar in appearance may have very different meanings in reality. Thus the word *income* may have different connotations depending on whether the income is actual or imputed per unit area of woodland. And a classic case of a word meaning 'what I choose it to mean', to use Humpty Dumpty's dictum, is provided by the way in which capital gains tax in the USA came to be adjusted between 1943 and 1986. Here, as Duerr (1960) explains, redefinition of what constituted capital gains and which therefore benefitted from the concessionary rates of tax applied to such 'gains' meant that most ordinary income from cutting timber ranked for capital gains treatment. An extreme example of the way in which words can mislead is the statement for Bulgaria in the FAO collection of forest policy statements (FAO, 1988) that 'Taxation policy is designed to facilitate forestry operations'. In 1988 all Bulgarian forests were public and the implication of the statement is simply that money arising from wood sales by lower level units is remitted to the headquarters office of the state forest service for redistribution on approved silvicultural operations: hardly a revolutionary concept and certainly having nothing to do with taxation as the word is normally interpreted. Finally, it should be noted that in several languages the term 'taxation' refers not to the raising of money by the state or other authority but the process of making an assessment of the property for such purpose; the word has subsequently taken on the meaning of survey.

Types of tax

Conventionally, a distinction is drawn between direct and indirect taxation, with the former implying charges on the person aimed at his income or wealth, and the latter aimed at his expenditure, largely but not exclusively on consumption, with payment of tax made automatically in the purchase of goods or services. Increasingly, the basis of such taxation is on value added by the producer of the good, the inputs to his production process having in turn borne their slices of VAT. The amounts of tax raised by direct and indirect taxation differ over time and among countries. It is reasonable to expect that the balance may alter because of changes in attitude towards the desirable distribution of (post-tax) income, but it has to be recognized that membership of the European Community means that the convergence of indirect taxation, as recorded in Appendix II, is marked and continuing. VAT is ignored in the discussion of tax burdens in individual country chapters.

In addition, the purposes of indirect taxation may change. Thus from 1946 to 1990, the French government operated a system of taxes on wood products – at least a selection of such products – as the sole means of financing its Fonds Forestier National. The European Commission objected that the levy discriminated against consumers of wood products and therefore represented a distortion of trade. Curiously, the revised basis (Ministère de l'Agriculture, 1991) still runs counter to this requirement. In another field altogether, planned discrimination is the intention of the carbon tax under discussion in the Community and several other countries. The 1991 proposal envisages a levy on industrial consumption of fossil fuels and emission of carbon dioxide to the atmosphere: it would be positively discriminatory against such activities. (It may be noted that similar taxation of animals, such as grazing animals, emitting methane would be easily achieved on the basis of good statistics, but because emissions from certain land uses are more difficult to assess, discrimination among farmers would result if an equitable method of assessment could not be agreed for these other, soil, sources.

The forms of direct tax

Direct taxes may be levied on income, wealth and transfers of wealth. Very different meanings may be ascribed to these words in the tax systems of different countries.

Income tax

Income is here taken to mean either actual income or imputed income. Whether actual income has to be aggregated with other income or not is an issue of crucial importance to the taxpayer. On the one hand splitting income, as is now possible for married couples in many countries, obviously reduces the total tax liability. On the other hand the ability to combine incomes implies that losses in one reduce the total sum of income on which tax is payable.

An essential difference exists between countries in the ways in which income is assessed, in particular whether as actual cash receipts or as calculated annual average expected income assuming a sustained yield unit. In Britain the position under Schedule D of income tax until 1988 was, and in some European countries remains, that the charge to tax arises from actual net income, that is after allowable expenses have been deducted. But in other countries, such as Belgium, France and Germany, a standardized notional income is assumed usually equivalent to the average annual income expected to result from conventional management of the site in question over the crop's life.

The method of assessment using calculated income carries several implications, notably that the bureaucratic burden of reassessment as prices, etc. alter is not to be ignored. There will generally be a tendency to delay revision, and tax revenues will then fall because of inflation. In Germany, for example, there has been no change in the basis of valuation for both income and capital (land) tax on woodlands between 1935 and the present although a review was conducted in 1964. Also, values ascribed must be set conservatively in order not to penalize the marginal taxpayer for whom, for example, average wood prices may represent unrealistically high values in his particular locality. The tax tends in fact to become one not on income in the conventional sense but on one form of wealth, namely property in the form of woodland.

Wealth tax

The bulk of OECD countries have net wealth taxes (OECD, 1988), with reliefs depending on the form in which wealth is held, with business assets, in contrast to other forms of property, often constituting an exempt or partially exempt class. Table 3.1 shows the sums of wealth and inheritance tax receipts as a proportion of GDP in the countries reviewed in this study. It is almost certainly true that the percentages for the private forestry sector alone are higher but data are lacking for both numerator and denominator in all countries.

Table 3.1. Receipts of wealth[a] and inheritance taxes as a proportion of gross domestic products (GDP).

Country	Percentage	Country	Percentage
Switzerland	0.98	Norway	0.29
Denmark	0.45	Belgium	0.27
The Netherlands	0.42	UK	0.24
France	0.39	Germany	0.16
Sweden	0.34	Ireland	0.12

[a] All countries listed except Belgium, Ireland and UK had wealth taxes at the time of the OECD survey.
Source: OECD, 1988, Table 5.2.

Property (land) tax

Taxes specifically levied on real property, in particular land and buildings, are common. The terms adopted to describe different forms of land tax are legion and reflect the differing purposes for which the tax was first introduced and the different authorities that raise revenue from them. The chapters on individual countries set out the bases used to assess land value and to set tax rates.

Capital gains tax

Capital gains taxes are common but their implications for the wood-producing function are liable to be so serious that almost all countries devise systems of reducing or eliminating their effect. Usually this is by relieving the growing stock of any liability. Perhaps this is the one point at which the familiar watchword that trees are both factory and product has its most direct relevance to a real-life practical issue.

Capital transfer, including inheritance, tax

Finally, capital transfer charges, covering gifts *inter vivos* and bequests, are important for PF owners, largely because the value of the capital employed in a forest is so large relative to income and hence to the capacity of the beneficiary to pay even though the charge falls infrequently. Along with other capital taxes, notably land (or property) tax, capital transfer tax lies at the root of the questions of tax neutrality, resource values and the extent to which non-market benefits are capitalized into asset values. Partly because of the capital-intensive aspect, a very wide variety of devices is deployed to minimize the burden on forests, to discourage fragmentation and to encourage good management. Among western European countries, only Italy and Britain have estate duty type taxes, i.e. taxes dependent on the size and composition of the estate, in contrast to inheritance-type taxes under which the charge bears on each beneficiary, with different schedules of tax rates applying to the different degrees of consanguinity.

Other general points on tax

The general point has already been noted that like the weather, a change in one part of the system necessarily has effects on different parts of the world sometimes far distant from the original disturbance. The flow of information on tax matters is highly organized and portfolio managers and those managing large estates are particularly responsive to changes. Indeed it is an interesting question as to whether forestry gains or loses from the narrow institutional frame within which forests are held and managed in many countries. Most private forests are owned by individuals, and unlike forest products firms, firms owning forests only are the exception rather than the rule in most countries.

Effects of tax arrangements

The forestry literature is littered with assertions about the effects of particular tax regimes, and none more than the former Schedule D/B arrangement of British income tax (see, for an example, Council for the Protection of Rural England, 1986). However, convincing demonstrations of tax effects are hard to establish. In the British case the complaint largely relates to the large scale of the afforestation operations, to the fact that these were conducted by contractors drawing labour, it is claimed, from many miles away and to the supposition that the rich absentee owners had little interest in the impact on the locality. Some of these detractions can hardly be ascribed to the tax system itself.

Most countries make very favourable provisions for afforestation and the gross effects, measured in terms of hectares established, are clear for all to see. Finer differences, such as the species used or the layout adopted, or management of the growing stock of established forests, are admittedly less easily controlled through the tax route. Indeed the position in relation to management of the growing stock is often so much less sharply focused that even in countries with large and long-established forestry traditions, the impacts of tax are not well understood. Carlén (1986) and Aronsson (1987) studied the impact of income taxation on cut of NIPF in Sweden. Despite deploying high-powered econometrics they were unable to show whether a progressive tax system increased cut or not. Their failure suggests that prognosis of the effects of alterations in taxes and, *a fortiori*, the imposition of a new tax have little chance of success. Sometimes the results of taxes may be positively detrimental judged by conventional forest management standards. For example, as seen with the former Estate Duty provisions in Britain, acquisitions and sales of appropriate-sized portions of forests can become an essential tool in capital tax planning with no relevance to effective physical management of the property. The position is exacerbated where, as in Scandinavian countries, heavy wealth taxes are levied, for capital unlike income provides the convenient means for borrowing against assets and so reducing net wealth.

Consideration of tax in the present study

So far as practicable, a numerical assessment of the direct tax burden of an average-sized individually owned private forest is made in the succeeding country chapters. The intention is to attempt to show the broad level of tax burden prevailing in different countries and the relative importance of the different taxes. The calculations are therefore stylized since the kinds of manipulation that arise cannot be summarized in the results of a single, formal calculation. In addition it is almost impossible with the information available to make a defensible numerical assessment of the total value of concessions that governments make to private forestry, although indications of the value of these are given where practicable. The results must, therefore, be treated with the greatest caution.

Grants

In contrast to taxes, information on grant levels, uptake and impact is much more readily available (see Johnson and Nicholls, 1991, for a clear summary of Forestry Commission and other schemes operative up to 1991 and Chapter 5 for Forestry Commission schemes at 1 April 1992). This is largely because there is not the same sensitivity and hence confidentiality surrounding the matter of such payments. In the main there has not been so much public debate on the subject of grant structures as there has been of taxes. There has been much discussion of arrangements in Britain in recent years of the frequency of alteration of schemes and the minutiae of the rules. A criterion of a good game is that the rules are simple: chess, for example, stands that test extraordinarily well. The main consideration noted in this study relates to the contrasting attitude of countries to the use of grants determined as a rate per cent of approved expenditure as opposed to the use of absolute levels of grant.

Loans are ignored in this discussion. Their administration creates major problems and the costs arising appear not worth any slight reduction in discounted public expenditure that might arise from their use in place of grants.

Research and Extension

Large sums of public expenditure are devoted to research carried out directly by the State or on contract to private or quasi-private organizations. The value of this work is widely trumpeted, in general with justification although it is easy to identify some projects which excel in their contribution. Whole research programmes are never evaluated in money terms. It is clearly important to consider how the balance of research and financial aid directly delivered to the owner through tax relief or grants should be struck. It is impossible to divorce the question from that of expenditure on extension. Objective scientific research, that is research directed to particular objectives, is pointless and applied research valueless unless it is applied, but the means of getting the message to field operators and managers is far too rarely considered. The views of Adams (1991) find echoes in Europe. He points out that in the USA the widening range of issues and diversity of owners calls for ever greater coordination of research, education and assistance. These crucial connections are discussed in the final chapters.

References

Adams, P.W. (1991) Addressing forestry issues for the future: a vision for coordinated research, education and assistance programs. Voluntary paper, 10th World Forestry Congress, Paris.

Aronsson, T. (1987) Forest taxation, neutrality and roundwood supply – an econometric analysis. Swedish Agricultural University, *Department of Forest Economics Report 76*. Umeå.

Byron, N. and Boutland, A. (1987) Rethinking private forestry in Australia – 1. Strategies to promote private timber production. *Australian Forestry*, 50, 236–244.

Carlén, U. (1986) The Swedish private non-industrial forest owner's cutting behaviour – an econometric study based on cross-section data. Swedish Agricultural University, *Department of Forest Economics Report 66*. Umeå.

Council for the Protection of Rural England (1986) *Written evidence to the Select Committee on the European Communities: EEC Forestry Policy*. House of Lords, Session 1895–86, 24th Report. HMSO, London.

Duerr, W.A. (1980) *Fundamentals of Forestry Economics*. McGraw-Hill, New York.

Ellefson, P.V. and Lyons, J.R. (1989) Policy development and program administration. In: Ellefson, P.V. (ed.), *Forest Resource Economics and Policy Research*. Westview Press, Boulder, pp. 228–240.

FAO (1988) Forestry policies in Europe. *FAO Forestry Paper* 86, Rome.

Firn Crichton Roberts Ltd (1990) *The Attitude of Landowners to the Woodland Grant Scheme: a Report to the Forestry Commission*. Pittenweem, Fife.

Forestry Commission (1974) *53rd Annual Report and Accounts 1972–73*, 8–9. HMSO, London.

Forestry Commission (1990) *70th Annual Report and Accounts*. HMSO, London.

Greene, J.L. and Blatner, K.A. (1986) Identifying woodland characteristics associated with timber management. *Forest Science 32*, 135–146.

Hart, C. (1979) Effect of taxation on forest management and roundwood supply in Europe. Report to the Joint FAO/ECE Working Party on forest economics and statistics and the European Forestry Commission.

Johnson, J.A. (1990) The social framework of market failure and intervention in forestry and environmental issues. In: Whitby, M.C. and Dawson, P.J. (eds), *Land Use for Agriculture, Forestry and Rural Development*. Proceedings 20th. Symposium of the European Association of Agricultural Economists.

Johnson, J.A. and Nicholls, D.C. (1991) The impact of government intervention on private forest management in England and Wales. *Occasional Paper 30*. Forestry Commission, Edinburgh.

Kay, J.A. (1989) Research and policy: the IFS (Institute of Fiscal Studies) experience. *Policy Studies* 9, 20–26.

Klemperer, W.D. (1989) Taxation of forest products and forest resources. In: Ellefson, P.V. (ed.), *Forest Resource Economics and Policy Research: Strategic Directions for the Future*. Westview Press, Boulder.

Kurtz, W.B. and Lewis, B.J. (1981) Decision making framework for nonindustrial private forest owners: an application in the Missouri Ozarks. *Journal of Forestry* 79, 285–288.

Ministère de l'Agriculture et de la Forêt (1987) [Private forest estates.] *Collections de statistique agricole; Étude* no. 268, Paris.

Ministère de l'Agriculture et de la Forêt (1991) [The reform of the levy on forest products.] *La Forêt Privée* 33(196), 58–61.

Normandin, D. (1987) [Management of private forest enterprises; structure and activities.] *Revre Forestière Française* 39, 393–407.

OECD (1988) *The Taxation of Net Wealth, Capital Transfers and Capital Gains of Individuals*. OECD, Paris.

Riihinen, P. (1986) Future challenges of forest policy analysis. In: Tikkanen, I. (ed.), Analysis and evaluation of public forest policies, *Silva Fennica* 20, 260–264.

Tikkanen, I. and Vehkamäki, S. (1990) *The Effects of Economic and Forest Policy on Roundwood Supply*. Proceedings Division 4, XIX Congress, IUFRO, Montreal.

Yoho, J.G. (1985) Continuing investments in forestry: private investment strategies. In: *Investments in Forestry Resources: Land Use and Public Policy*. Westview Press, Boulder and London, pp. 151–166.

Young, R.A. and Reichenbach, M.R. (1987) Factors influencing the timber harvest intentions of nonindustrial forest owners. *Forest Science* 33, 381–393.

4

Private Forestry in a European Context

Countries and Institutions

When Anderson (1950) wrote his memoir on the State control of private forestry under European democracies, countries had hardly recovered from the Second World War, gross world product had not yet regained its pre-war peak level, countries had only just reformulated their forestry goals and reshaped their forestry administrations, the EEC was hardly a gleam in Henri Spaak's eye and the European Free Trade Area had not had to be invented, while a new Soviet empire reached into a partitioned Germany. Geo-political goals were paraded as imperatives and powerful military alliances formed and, under aggressive leadership from the United States in the pursuit of more rapid economic growth, the General Agreement on Tariffs and Trade (GATT) was established. But, this apart, international obligations in the economic sphere were very limited, and in environmental or conservation fields such obligations were unheard of and largely undreamed of. Since that time five developments with major implications for forestry have occurred. First, there has been the longest period of sustained economic growth ever, with massive effects on the pattern as well as the total scale of demand. Second, there has been a significant development of concern with the state of the countryside environment, and national laws, as well as regional and local arrangements for physical planning, increasingly reflect this. Third, there has been a major growth in international treaty obligations which affect the daily lives of people in ways and to degrees never experienced in sovereign countries in earlier times. Fourth, and with substantial implications for shifts in supply of forest products in the short term and for demand in the medium and long term, there has been the dissolution of the Soviet empire. Finally, and arguably less significantly in the long sweep of history, there has been a reaction to the post-war devotion to social democratic ideals which has had dramatic effects on attitudes to the role of the state and dependence on the market.

Despite some of these influences and on account of others, there has been an increase in control of private activity in many walks of life. Forestry is no exception. Classically, forest influences in the sense of protection of the landscape and settlements were prayed in aid by foresters as justification for state intervention (Troup, 1938). With the growth of demand for wood products and subject to the tacit aim that prices should not be allowed to rise in real terms, western European governments have been uniformly intent on undertaking major programmes to secure larger wood supplies. The private sector played a notable part in this in all democratic European countries apart from Spain, Ireland and, until the 1970s, Britain. In the United States a prospective wood shortage was, as so often in the past (see, for a clear-sighted review of the impact of changing opinions on timber primacy, Bennett, 1968), paraded as the bogeyman justifying programmes of assistance to the non-industrial private forest owner. In Japan, the reconstruction of seriously depleted forests following the war has been pursued principally through incentives to the private grower (Handa, 1988).

In considering private forest policy, the extent of private forests in continental Europe, their long traditions of intensive silviculture and management and their apparent success in adjusting to the massive changes in the social and economic environment during the past half century make European countries an obvious source of guidance. Other countries with similar cultural and political backgrounds to Britain's are also of interest.

Before reviewing elements of policy or the implementation of policy in individual countries, some attention is paid to international institutions. These are, however, not of central concern to the description and assessment of policy towards private forest owners, but are considered because they may influence many of the countries surveyed on particular matters, one obvious example being the afforestation of setaside arable land.

The European Community

Forestry connections

The following Directorates General (DG) have some specific bearing on forestry:

III – *Industry*: deals with wood as a raw material for traditional processing industries.

VI – *Agriculture*: covers agriculture which is still the dominant force

[1] A useful, if slightly dated, compendium of facts and benchmark events in the Community's work on forestry appears in CEC (1986). A statement of the strategy of the Forestry Action Plan appears in CEC (1988a).

controlling afforestation rates and locations through the economic availability of land, and forestry[1].

V – *Social Affairs*: slight contact only, deals with employment aspects.

VII – *Development*: a misleading title since this DG only covers African, Caribbean and Pacific countries, but deals with the Lomé Convention on trade in tropical products and Tropical Forestry Action Plans.

XI – *Environment*: environmental aspects of forestry programmes are covered as well as coordination of EC policies on tropical forestry.

XII – *Research*: has covered aspects of forestry such as wood conversion and utilization, and pollution effects on trees;

XVI – *Regional Policy*: covers promotion of industry, and forest development in association with industrial development;

XVII – *Energy*: covers wood as a fuel.

The influence of different parts of the European Commission varies with time and the policy issues of the moment. In the medium term that of Agriculture must be rated highest, owing to the rapid changes in prospect as a result of the policy adjustments in agriculture which include the use of tree growing as a palliative. Environment, subject to developments on subsidiarity, may have a strong effect through the application of environmental impact studies and the adoption of rules on factors such as conservation effects of particular forestry practices, as well as outside through regulations on water quality, pollution emissions, etc. On the whole it seems likely that this function will continue to bear, and probably bear increasingly, on forestry. The impacts of Community policy on carbon taxation could be far reaching.

The activities of Energy, Research and, through policy on trade in tropical wood and its impacts on the price of temperate wood generally, of Development and Environment may be expected to have some, but smaller, influence.

The Forestry Action Programme

This programme, established through Council Regulations nos. 1609 to 1615 inclusive, has generated the programme with the widest and largest effects on the forestry field thus far. It is useful to put spending planned under this programme into perspective. Total EC expenditure in 1991 was planned to amount to 55 billion ECU (CEC, 1990). Of this, 35 billion ECU was to be allotted to the agricultural structural and guidance funds and much of the Regional and Social Funds spending is also devoted to agriculture, implying that this sector would receive some 84% of the Community's total budget or 46 billion ECU. Spending in 1991 under the Forestry Action Programme launched in 1989 was planned to amount to 75 million ECU.

Regulation 1609/89 on afforestation of agricultural land provided for reimbursement of any person or firm except state forest services at specified rates for planting, roading and other afforestation related operations. An annual 'premium' is also payable to the occupier for up to 20 years. Other Regulations are not of such importance or relevance to the United Kingdom as this. No. 1610/89 entitled 'Develop and optimise woodlands in rural areas' applies to certain low income rural areas and allows for percentage reimbursement of costs of a wide range of forestry operations. Other measures include that on 'Protection of the Community's forests against atmospheric pollution', Regulation 1613/89, which is based on No. 3528/86 and provides for monitoring of forest condition, research, trials and demonstration projects. This is regarded as particularly important in Germany. Similarly Regulation 1614/89 dealing with protection against fire builds on an earlier Regulation, 3529/86, and provides both higher rates of assistance and substantially larger total sums. Its main beneficiaries are countries bordering the Mediterranean. The overall programme expenditure projected under the seven Regulations was some 230m ECU over the 4 years 1989–1992.

1992: The European Single Market

The Single European Act (SEA), signed by Heads of State in 1986 and ratified in the UK in 1987, made the first amendments to the Treaty of Rome since the treaty was established. In addition to the objectives of an increasingly controversial character, at least as viewed in Britain, concerning economic integration, with economic and monetary union as main planks, SEA incorporates two aims that are highly relevant to forestry. One is the recognition of the desire to work towards the protection of the environment at large as well as in the workplace. This feature emerged with the major growth in concern over environmental matters in the 1970s and 1980s. The relevance to forestry lies in the incorporation into EC legislation of certain 'environmental' aims. These are very extensive and include changes of land use such as afforestation. As a result of the debate concerning the Maastricht Treaty, it is conceivable that changing interpretations of subjects to which subsidiarity applies will have important implications for essentially local environmental issues.

The other element of importance to forestry was the impetus the Act gave to progress towards the single unified market, including the removal of physical, technical and certain fiscal barriers to trade. The target date set for completion of measures needed to establish the internal market was the end of 1992, though this particular date has no legal force and in practice many component parts could not be achieved by that date.

Appendix II sets out the economic arguments for further integration and the steps being undertaken which have a bearing on forestry. The main subject-related matters that appear likely to influence forestry in the medium term relate to plant health regulations and timber standards. The latter may allow increased access to the British market and, as a result, some further downward pressure on wood price.

Research

Substantial expenditures have been devoted to research on wood growing and processing in the two Programmes entitled 'Wood as a renewable raw material' (WARRM), with 12 m ECU over the period of the first, 1982–1985, 7.5m ECU in the second, 1986–1989 and 12m ECU in the FOREST programme 1990–1992. Like those of other Community programmes, one of the principal objectives has been to organize collaboration in mutually profitable projects by workers in different Member States. The 1993–1997 'Agriculture and agro-industry research' programme includes provision for, but no specific allocation of funds to, forestry research.

The FAST (Forecasting and Assessment in Science and Technology) programme of 1984–1987 is mentioned here for completeness and because it contained a forestry component. Its aims were theoretical. Some notion of the programme's concern in relation to forests may be judged by the title of the subprogramme in which forestry fell, namely the 'Integrated development of renewable natural resources'. The report reviewing work (CEC, 1988b) justified the interest in forestry on the grounds of the 'poor integration between wood production and processing, prospective world "excess" demand potentially sending prices soaring, a concern over higher value products being imported and the need to improve the competitive potential in relation both to wood supply and processing.' Although one can agree with the last point, others display an uncritical acceptance of the familiar cry about imports. Curiously little mention was made in the reports of environmental aspects apart from air pollution damage. It is unfortunate that the work was technically rather than economically orientated since the questions on the efficiency of supply, etc. deserve more attention. The study made little discernible impact either on forest or forest research policy.

Tropical forests and tropical timber supply

The European Commission has taken a position on tropical forestry in its Communication on the conservation of tropical forests (CEC, 1989). This contains proposals for tropical timber marketing, an area in which commercial interests in the timber trade have wielded considerable influence. There can hardly be any doubt that the bulk of tropical forest logging is not followed by sustained management, but equally that without a complete trade embargo by importers, the exporting nations will continue to earn foreign exchange and provide domestic income through profits and labour earnings. The relevance to European growers of any fall in tropical imports lies in the increase in demand for the corresponding specifications. By no means all of these are for high density ornamental hardwoods. Some enter western countries as bland sawnwood for which clear softwoods would do duty, others as plywood for which birch and Douglas fir ply or other panel products serve the same role. Producers including growers in Europe therefore stand to gain over a range of species and products from restrictions on tropical wood trade.

Carbon tax

Industrial pressures again come to the fore in the discussion of a Community proposal for a carbon tax (CEC, 1991). The proposed basis for this is a combined tax on energy production and on CO_2 emissions.

EUROSTAT

The Statistical Office of the European Communities (SOEC) at Luxembourg must be mentioned for its work in the field of income statistics. While the provision of harmonized statistics on agricultural incomes is clearly essential for the purposes of agricultural policy and is achieved, the Office has not been able to persuade countries to improve their data collection in the field of income derived from forestry. The latest publication (Eurostat, 1990) indicates the incompleteness of the statistics available for the sector. Thus for nine countries, excluding Ireland, The Netherlands and Portugal, only six could provide estimates of gross fixed capital formation by the sector and only five net value added including data on employees' remuneration. It should also be noted that an attempt to prepare data on economic measures of pluriactivity, that embraces non-agricultural activities on farms, has not been met. Data in this field would be valuable in relation to Community policy towards farmers with small woodlots who might be covered by the rural areas part of the Forestry Action Programme or by the long awaited policy on agricultural adjustment.

SOEC compiles some data on size distribution of forest holdings, as well as areas and wood production by species. This information is more detailed than that collected by ECE/FAO on forest areas, growing stock and other physical measures of forestry stock and output, but the latter source is to be regarded as the main source of such information.

Although Eurostat has compiled wood price data in the past, the principal initiative in this area lies with FAO, Rome (see FAO, 1990 for an example of the output of the regular programme of price statistics for roundwood and many forms of processed wood) and with ECE/FAO in a developing programme concentrating on Europe with more detail on price structures (ECE/FAO, 1989).

Other Relevant International Institutions

Comité Central de la Propriété de la CEE (CCPF)

The CCPF aims to represent the interests of non-State forest owners in Community countries. Its activity is limited but steps were under way in 1991–1992 to stimulate the organization and develop its functions. The principal intention of a policy-oriented group (Dansk Skovforening, 1991) was to work out appropriate relations with the European Commission and other EC organizations, including information channels through which it might best exert influence.

Confederation of the European Paper Industry (CEPI)

This organization is pan-European and replaces the former CEPAC, which had a Community base. Some indication of its active approach and the value of strong industry support in preparing and presenting its case may be gained from CEPI (1992). This document provides a clear analysis of forest industry handicaps and action required by the various authorities to improve the future supply of wood. Its list of recommended measures is so wide ranging that it is set out in abbreviated form below:

1. European Commission to coordinate existing official forestry bodies at Community level.
2. The Commission to implement action based on actual wood requirements and try to disengage forestry from the limitations of the Common Agricultural Policy.
3. The Community and member States to improve the fiscal and legal basis for private forestry.
4. The Commission to identify geographical areas for forest expansion.
5. The Commission to analyse forest ownership structures particularly in relation to their impact on wood supply.
6. The Commission to intensify support of cooperatives.
7. Public sector, including EC, funds for research to be increased.

The CEPI assessment and suggested remedies add up to a coherent approach by a highly interested lobby. The views expressed are a useful and cogent contribution to the debate on forestry which has become increasingly dominated by claims for ever greater provision of non-market benefits.

Confédération Européenne de l'Agriculture (CEA)

The Confederation is also pan-European. Its annual deliberations (CEA, 1991) include sessions on forestry which serve as valuable reminders of the policy, including implementation, issues which are of current concern to owners of larger private forests in particular.

Economic Commission for Europe

This United Nations body carries out several very important functions serving forestry and the wood-using industry. Those most relevant to this study are the preparation of periodic studies of timber trends (for example ECE/FAO, 1986), the compilation of country and regional forest resource assessments, and international exchanges on technical matters, especially concerning forestry operations. Selected data from the 1992 compilation on forest resources are shown in Appendix I.

In addition ECE, through its Europe-wide membership is responsible for activities under the Convention on long-range transboundary pollution. These include the organization of forest surveys (ECE, 1988) and specific tasks, such as forest increment studies, which seek to improve knowledge of forest condition. Some of the reports reviewing the results of tree 'health' surveys suffered from an unjustifiable willingness to interpret any crown thinness or leaf discoloration as evidence of air pollution damage.

General Comment on Statistics

The basis of government intervention in any sector should always be an objective appraisal of the situation. This appears too obvious to state but experience teaches how major decisions on programmes and taxes are taken without the necessary factual foundation. Statistics are also necessary in order to monitor whether a given programme has had the desired effect. Finally one would expect that those active in the sector itself would require sufficient data to be able to keep a finger on the pulse of the industry.

The forestry statistics given in Appendix I refer to physical features. Traditional forest statistics such as the scale of a country's forest estate, the size of the wood-based industries, or the size of the annual programme of new planting are not necessarily the facts that are particularly relevant in the consideration of a country's forest policy. Policy matters relate to far deeper concerns than a few statistics on the physical scale of forestry and forest-based industry can reveal. Much more may be required on aspects such as levels and trends of logging costs, the place of forests in the recreation resources of the country, or the size distribution of private woodland holdings. Steps are being taken by ECE/FAO's Timber Section to extend the reach of data to include wider physical measures as well as some social material. In addition much more needs to be done to compile economic statistics such as national income generated in private and public forestry, and the regional distribution of forestry employment in private and public enterprises.

The Choice of Countries Surveyed

This study is not intended to be comprehensive in its coverage of western European countries. In considering which countries to observe, it is clearly right to keep an open mind about the places in which interesting and potentially relevant developments are occurring. No country, now that centrally planned economies are no longer fashionable, can be dismissed. One reason for concentrating attention on western European countries is that their generally similar cultural traditions imply that the same trends may appear in Britain. Any common strands within countries that are members of the European Community are important because they may well become incorporated into Community policy or practice. While it has thus been considered that countries of north western Europe which share a similar cultural attitude to society and land use make these the most appropriate to note, whether or not they are members of the European Community, other countries such as Australia and New Zealand are relevant. Comments on the private forestry situation in these countries are therefore included (Chapter 15), together with notes on the changing situation in eastern Europe because of developments on privatization there. Finally, because of the importance of non-industrial private forests in the United States and the attention given in that country to studies of the effectiveness of public policy, the commentary in Appendix III on policy analysis with special reference to extension services emphasizes the position in the United States.

Subject Coverage of Country Chapters

Country chapters summarize current aims of forest policy[1] and implementation measures concerning private woodlands. Their principal purpose is to state the factual position as of early 1992 on these points for each of the countries reviewed. Reference is also made to issues which appear to

---

[1] Richards (1987) gives a useful description of forestry policy and statistics, and forest products industry in ECE countries.

be of particular interest to the outside observer. It has been considered, however, that there is little point in describing current government procedures for policy formulation or the organization of the administration of private forestry except in the broadest outline. The arrangements for policy review depend heavily on national conventions whereas the allocation of responsibilities for forestry changes too frequently and makes information on this point too transient (Prats Llaurado, 1991 gives a useful survey of the worldwide position on forestry administration). However, organizational aspects are recorded where considered relevant, for example because others' arrangements may provide useful pointers.

Public Financial Support

Data on financial support are believed to be correct as of early 1992. It has been considered pointless to attempt to make any economic calculations on the prospective returns from private investment in new or existing woodlands. Compiling the relevant representative costs, yields and wood prices alone would make this a daunting task, and in addition variations in land tenure, the personal circumstances of owners and their taxation position and also the various grant arrangements which are often highly differentiated as to species or location make it impracticable to estimate returns to capital in any fully satisfactory way. Even the detailed assessment of afforestation in Germany by Plochmann and Thoroe (1991) fails to cover tax aspects. Instead, the succeeding chapters on west European countries, except that for Germany where each Land has its own set of grants, report forestry grants in detail in order to indicate the complexity of grant arrangements.

Illustrative tax burdens have been assessed for an average woodland holding in each of the western European countries visited. The purposes are twofold, first to give an approximate measure of the total charge on a private individual's holding after reliefs are incorporated, and second, to indicate the relative importance of different taxes. In calculating the incidence of the various taxes, an average size of holding has been assumed and estimates implicitly made of the owner's other income and assets. Recent levels of cut and hence income as well as costs of other operations have been assumed although in some countries cut in recent years has been below that consistent with sustained yield. Annualized sums have been calculated for intermittent taxes on capital (inheritance). The convention adopted is to include the value of the annual payment which would have to be set aside to provide for the lump sum tax arising at the end of the period of an individual's ownership and payable at death, on a gift or as capital gains tax on sale of forest. This method of assessing the annual equivalent levy happens to be consistent with the general practice adopted in inheritance tax planning in the United Kingdom, namely payment for an insurance policy during the period of ownership against the ultimate charge (for an example of the approach in the field of forestry, including tax relief on creation of an Inheritance Tax fund, see Orde-Powlett, 1990). The annual equivalent, disregarding income tax offsets, is

$$S(k - 1)/(k^n - 1)$$

where s = capital tax due, k = 1 plus the rate of interest in real terms, n = number of years the forest is held.

Based on a review of the long-term real pre-tax rate of return on financial assets in Denmark since 1819, Holten-Andersen (1991) suggests that this has averaged a little over 3%, and proposes a rate of 2% as an appropriate measure of the post-tax rate. The pre-tax rate in Britain for a comparable period has also been about 3%, implying a post-tax rate of about 2.5%. A value of k of 1.02 is adopted: with $n = 25$ this implies an annual multiplier of 0.0312, or, with $n = 30$, 0.0246.

The tax sums derived are inevitably crude. This is partly because it is not possible to make accurate allowance for the average impact of the various routes adopted by taxpayers to lighten the tax burden. In addition, no recognition is given in the calculations of the provisions made in many countries' forestry tax systems to deal with fluctuations in prices and/or yields from year to year.

Other Public Expenditure in Support of Private Forestry

Research

As far as possible, estimates of public expenditure on forestry research have been obtained from the appropriate authorities in countries. In 1990, the European Commission set up under COST (European Cooperation in the field of Science and Technical research) a group on forestry and non-wood products research which, among other activities, compiled data on person-years. The survey does not indicate the dependence of the programmes covered on public sources of finance. This is important in relation to the emphasis of the present study which is on public policy towards private forestry. Data on expenditure per research worker are very slender but figures provided (Pardos, personal communications) indicate that costs per person-year of scientists and technicians do not vary greatly among countries. The values for *c.* 1990 are: Denmark 54,900 ECU, Sweden (with a high proportion of technicians) 48,100 ECU, Switzerland 61,500 ECU. For Great Britain another source (FRCC, 1990) shows 52,000 ECU.

The allocation of research expenditure to private non-industrial forestry is necessarily arbitrary; each country chapter states the rationale for the choice of proportion.

Extension

Interest attaches to the dissemination of information gained from research and other sources. Wherever possible an estimate of public expenditure on extension is recorded. However, reliable data on this activity are difficult to obtain. First, for many forest service field staff, extension is one among many private forestry and even state forestry tasks. Second, few services require staff to account for their time in detail. Third, the definition of extension differs among countries; in many cases it includes education of others than forest managers and workers, and embraces parts of grant administration. The figures quoted should therefore be treated with caution.

References

Anderson, M.L. (1950) State control of private forestry under European democracies. *Oxford Forestry Memoir* no. 22. Clarendon Press, Oxford.

Bennett, J.D. (1968) *Economics and the Folklore of Forestry*. University Microfilms, Inc., Ann Arbor, Michigan.

CEA (1991) [43rd. General Assembly, Reports.] Brugge.

CEC (1986) *CAP Working Notes: special edition - European forestry*. (VI/BIB/SPEC.EDIT 01/86). Brussels.

CEC (1988a) *Community Strategy and Action Programme for the Forestry Sector*. COM(88) 255 final. Brussels.

CEC (1988b) *The FAST II Programme (1984–1987): Results and Recommendations*, Volume 6. Brussels.

CEC (1989) *The Conservation of Tropical Forests: the Role of the Community*. COM(89) 410. Brussels.

CEC (1990) *Preliminary Draft Budget for the EC for the Financial Year 1991*. COM (90) 1212-EN. Brussels.

CEC (1991) *A Community Strategy to Limit Carbon Dioxide Emissions and to Improve Energy Efficiency*. Commission Proposal 89/8/91. Brussels.

CEPI (1992) *Forest Resources and the Community Paper Industry*. Brussels.

Dansk Skovforening (1991) [Annual report.] Copenhagen.

ECE (United Nations Economic Commission for Europe) (1988) Forest damage and air pollution. *Report of the 1986 Damage Survey in Europe*. United Nations, Geneva.

ECE/FAO (1986) *European Timber Trends and Prospects to the Year 2000 and Beyond*. United Nations, New York.

ECE/FAO (1989) Roundwood price statistics and specifications, paper to the Joint working party on forest economics and statistics. Typescript, Geneva.

Eurostat (1990) *Economic Accounts for Agriculture and Forestry 1983–1988E*. SOEC, Luxembourg.

FAO (1990) Forest products prices. *FAO Forestry Paper* 95, Rome.

FRCC (Forestry Research Co-ordination Committee) (1990) Annual collation of forestry research - 1990. *FRCC Information Note* no. 27. Forestry Commission, Alice Holt Lodge, Farnham.

Handa, R. (ed.) (1988) *Forest Policy in Japan*. Association for research and publishing on Japanese forestry, Tokyo.

Holten-Andersen, P. (1991) [Inflation and taxation in the forest economy.] *Dansk Skovforenings Tidsskrift* 76, 232-300.

Orde-Powlett, R. (1990) Practical solutions to the 1988 budget. *Quarterly Journal of Forestry* 84, 118-120.

Plochmann, R. and Thoroe, C. (1991) [Support for afforestation.] *Schriftenreihe des Bundesministers für Erhnährung, Landwirtschaft und Forsten*. Reihe A, Heft 397. Landwirtschaftsverlag, Münster-Hiltrup.

Prats Llaurado, J. (1991) Forestry administration. Proceedings 10th World Forestry Congress, vol. 7; pp. 261-272. *Revue Forestière Française Hors série* No. 7, Nancy.

Richards, E.G. (ed.) (1987) Forestry and the forest industry: past and future. *Major Developments in the Forest and Forest Industry Sector Since 1947 in Europe, the USSR and North America*. Martinus Nijhoff, Dordrecht.

Troup, R.S. (1938) *Forestry and State Control*. Clarendon Press, Oxford.

5

BRITAIN

THE SCALE OF THE FORESTRY SECTOR

Measures of the scale of British forestry are few, firm data being confined to area, wood cut and employment generated. Unlike many countries, and despite the small size of the sector, Britain compiles data allowing the national income from Forestry Commission forests to be estimated, but no such computation is produced for private forestry. An important element of the assessment of national income in forestry is the valuation of net increment in the growing stock; this is a topic that must be attended to if Britain is to measure its success in moving to or maintaining a sustainable economy (Pearce *et al.*, 1989). It is ironical that valuation of stock change in countries' forests, such a vital element of income accounting of forests, has been most worked on in a few developing countries where forest depletion, as opposed to forest accretion, is the significant feature (Repetto, 1988).

Table 5.1 sets out the principal area features of Great Britain's woodland holdings in all forms of ownership. In British usage 'private' is the word used to identify all woodland owned or leased by parties other than the Forestry Commission.

A substantial proportion of the country's timber (28m m^3 relative to 197m m^3 in woodlands in 1980) occurs in hedgerow trees and spinneys. These are largely in private hands. Although forestry policy has by convention usually been concerned with woodlands of at least 0.25 ha and thus has not considered this category of tree, the resources are important by virtue of their place in the landscape. Because of their pervasiveness and this landscape role they remain an important determinant of people's conception of which trees are regarded as appropriate to the British, and especially the English, scene.

Table 5.1. Area of woodland by ownership and main forest type,[a] 1991 (000 hectares).

	High Forest		Coppice and coppice with standards	Other woodland[b]	Total all types
	Conifer	Broadleaved			
Private woodlands	710	532	39	162	1443
Forestry Commission	809	50	—	34	893
Total	1519	582	39	196	2336
Percentage 'private'	47	91	100	83	62

[a] for Great Britain only. Woodlands in Northern Ireland, for which the Department of Agriculture Northern Ireland is the responsible government body, are predominantly state owned and amount to 73,000 ha.
[b] 'other woodland', according to *Forestry Facts and Figures*, consists of 'areas where timber production is not a main objective. It includes areas managed chiefly for amenity and public recreation.' With the new roles required of woodlands managed with the aid of the Forestry Commission's Woodland Grants Scheme, sharp distinctions of the objectives served by different woods are increasingly difficult. The category 'other' however normally contains woods regarded by timber producers as scrub, including overgrown coppice.
Source: *Forestry Facts and Figures 1990–91*, Forestry Commission (1991a).

The Growth of Woodland Area

Woodlands cover 10% of Great Britain. This statistic (Table 5.2) has risen steadily from a value of 5.6% in the immediate aftermath of the First World War. The previously low value was a result of the dominance of agriculture, including upland grazing, as a form of land use for centuries past in Britain. (A useful survey of developments in forest history, together with a full bibliography, is provided by Miller, 1991.)

The proportion of land under woodland is considered by many a barometer of the success of government policy on the ground that a low proportion reflects a pattern of land use which is less than desirable, reflecting much 'under-used' land. It is true that the near doubling of land area under trees since 1919 has occurred largely as a result of afforestation in the uplands, especially in Scotland during the past three decades. It is difficult, and probably barren to attempt, to determine the contribution that the use of land as opposed to the wood supply argument bears in the discussion of the appropriate area of woodland for the country.

The Extent of Self-Sufficiency in Wood

The low level of woodland area as a proportion of the land surface compared with that in other European countries implies a low absolute level of wood production and, because Britain's wood requirements are substantial, a low

Table 5.2. Areas of woodland (million ha) by ownership in Great Britain at various dates[a].

	1913–1914	1924	1938–1939	1947–1949	1965	1980	1990
Private woodlands	1.27	1.20	1.20	1.20	1.09	1.22	1.44
Forestry Commission[b]	0.03	0.05	0.18	0.25	0.65	0.89	0.89
Percentage private	98	96	87	82	62	58	62
Total							
> 0.4 ha	1.3	1.25	1.38	1.46			
> 0.25 ha				1.48	1.74	2.11	2.34

[a] The two minimum size criteria imply that totals shown in the upper part of the table for earlier years are not directly comparable with those for 1965 or for 1980 and 1990.

[b] Crown office of Woods and Forests before 1919.

Sources: areas of woodland over 0.4 ha in extent (1 acre) calculated by Locke (1971) and over 0.25 ha from the 1979–82 census (Forestry Commission, undated) for 1980 and Forestry Commission (1991a) for 1990.

level of home production compared with total consumption. This ratio is a much-favoured measure in forest policy discussions. Unfortunately there is no consistent series showing the proportion that domestic roundwood production bears to total apparent consumption of wood in the rough since official statistics are confined to physical volume measures of wood consumption assessed in terms of wood raw material equivalent (WRME), that is before the impacts of recovery and re-use of wood fibre are counted. On the WRME measure, total apparent consumption was assessed at 48.5m m^3 in 1990 (Forestry Commission, 1991a). Since the equivalent volume of wood in the rough (EQ) is likely to be lower by some 15%, it follows that United Kingdom production of 6.73m m^3 over bark, that is including Northern Irish production of roundwood at 150,000 m^3, or 5.89m m^3 under bark amounted to some 14.3% of this measure of consumption in 1991.

Issues of the Current Policy Debate

Throughout the 20th century, policy on forestry has been founded on recognition of the high dependence on overseas supplies of wood products, usually assessed in its most dramatic form by referring to the value of net imports of wood, wood products, pulp, paper and board and their products. Play has also been made with the low output of marginal land and the small proportion of land under trees. Other considerations have included the extra employment claimed relative to other land uses, regional gains in income and jobs in rural areas arising in consequence, the downstream industries' labour demand and increased recreational opportunities claimed. The economic arguments have been the most worked over of all and these are therefore considered last in the following discussion.

The overwhelmingly important consideration in the minds of government, its forestry department, the Forestry Commission, and most foresters and woodland owners, however, remains the increase in home production of economically accessible wood. A constraining influence on this, over and above market limitations to the achievement of a high rate of new planting, has been the marked emphasis, steadily increasing in recent years, on the need to pay regard to a whole range of environmental considerations. These are diverse: they are also comprehensive since they relate to extension of the forest, to conversion of woodland from a seminatural form to one emphasizing commercial wood production, and to the management of other existing woodland.

The following paragraphs describe issues which appear important in relation to British forestry as a whole in the early 1990s. In effect they constitute an agenda for considering what elements British forest policy should address. Issues includes subjects, such as concern for the quality and appearance of the countryside, which must be ranked as objectives at the level of the debate implied by a discussion of national forestry policy, as well as topics such as the balance between private and public sectors which are concerned with means rather than ends. Chapter 16 extends the discussion of issues in the light of evidence from the countries under study.

The countryside environment

At one time, the word environment as applied to the countryside referred, broadly, to the appearance of the country, that is landscape. Separate bodies and societies dealt with river pollution, atmospheric pollution, noise and other nuisances. Today these influences are brought together under the one portmanteau word 'environment'. Because the public perception of what constitutes the environment covers not only the countryside but also the built environment, the atmosphere, the seas and indeed the whole of nature, the word has become invested with an importance that often exaggerates the real influence of a particular environmental 'good' or 'bad'. Nevertheless, there can be no question of the importance with which the word has become invested. Such issues as the pollution of land, water and atmosphere have come to exert great influence on governments as illustrated by the weight which politicians, often influenced by developments abroad such as lake acidification, have come to attach to them. In some countries parliamentary representation as Greens has obviously increased the leverage of those who espouse environmental causes. One governmental response to this rise in the importance attached to the environment has been the creation of environmental protection agencies. But beyond this a few significant political events, such as the publication of the Brundtland report, widespread support, even extending to the former British Prime Minister, Mrs Thatcher, in the late 1980s for actions to limit damage to the atmosphere and to mitigate climate change, and more recently the United Nations Conference on Environment and Development (UNCED) held in Rio de Janiero in 1992 stand out.

In relation to the rural scene a perceptive statement of the origins and growth of the environmental movement in Britain and elsewhere in Europe is given by Lowe (1987). He notes that the United Kingdom leads among European Community countries in its concern with rural amenity and conservation, and offers two principal reasons for this. One is the capital intensity of British agriculture which *ipso facto* has drawn attention to its effects and the rapidity of their appearance through the loss of species and habitats. The second is the small proportion of the population engaged in agriculture, implying a lack of ability on the part of the agricultural community to make its (economic) voice heard. Similar features characterize forestry. Countryside environmental groups can act as the focus of single interest groups in a way that might not be possible in countries where the links with jobs and income and the economic activities carried on are closer. These factors are in strong contrast to the position in countries like France and Ireland, where farming is regarded as the source of the peoples' spiritual and emotional roots. There is a marked tendency in Britain to value the countryside in aesthetic terms; rural economics are not considered. Lowe also points out that Nature is divorced from landscape. At the time he was writing, Britain was the only country with separate agencies for these two features, although this has now changed through the merger in 1990 of a dismembered Nature Conservancy with the three Countryside Commissions.

There are three dimensions which can be recognized, all of which concern the content of woodlands but with different emphases. One, landscape, in principle implies only visual access. This will often mean, however, that an interest will exist in gaining actual access to woodlands. The second, recreation, can be separated into general public access, the issue discussed below, and sports and other pastimes which can usually be treated as marketable outputs on which government has little to say. The third is conservation of wildlife. Landscape is currently not a continuously active matter although expressions of concern erupt from time to time. The 1988 decision of the Secretary of State for the Environment, whose remit on land planning matters ran only in England, provides a good example. He decided (Forestry Commission, 1989a, 1989b), in association with the forestry Minister for England, that large-scale planting of conifers above 800 feet (244 m) would be restricted. It should also be noted that an earlier and important policy initiative, in this case affecting forestry throughout Britain, had been taken with the announcement of a policy on broadleaved woodlands in 1984. Both these policy initiatives stem from strong feelings on landscape particularly, though with other concerns such as sporting and wildlife conservation playing their roles.

In Britain the third concern, wildlife conservation, has become the most important for both policy and practice. In recent years the most forceful, persistent and decisive criticism of the manner and rate of forest expansion and of treatment of existing woodland has stemmed from the nature conservation movement. The attacks made have been irrespective of the agents of change, whether public or private. Much of the criticism derives from a desire for the *status quo* to be maintained, that is to conserve habitats. This desire is often unsupported by knowledge of the means of conserving that are to be employed. In this discussion, a central issue is the scale of area to which claims for protection and maintenance of habitats and species should be extended. There is little argument about the desirability of maintaining certain areas as protected habitats. A thoroughly well-accepted expression of the aim is found in the nomination of National Nature Reserves. The lesser status of Sites of Special Scientific Importance, sites for which the more appropriate word 'conservation' rather than 'scientific' was originally proposed (Moore, 1987), offers another form of protection of the *status quo*. Presumably an absolutist position would require that the full range of diversity should be retained by appropriate designations of sites that are not completely dominated by anthropogenic influences. This attitude no doubt underlies many protagonists' arguments (Nature Conservancy Council, 1984, 1986; Ratcliffe, 1981, 1990) for severe restriction of afforestation, but the doctrine of primacy of one form of land use is not acceptable; the reality of the pressures to change land uses cannot be ignored.

International agencies, directly or indirectly, also influence British rural environment policy. These have been more active over the years, beginning with the 1972 Stockholm Conference which led in turn to the establishment of the United Nations Environmental Programme and from which the 1992 Rio UN Conference on Environment and Development sprang. The Organisation for Economic and Co-operation and Development also undertook a path-breaking role. Indeed it was in an OECD forum that the principle of the polluter pays was first agreed by governments. Finally, the European Community has become far more active in initiating legislation on environmental issues in recent years. In part this is because of the impact of the Chernobyl disaster which brought home the menace of cross-border incidents (Roney, 1990), in part because of the lead of certain countries, more sensitive than others to the excesses that were growingly apparent, in regulating new substances or actions capable of degrading the general environment (Arbuthnott and Edwards, 1989). A good example is provided by Germany's leadership in introducing lead-free petrol. To the extent that the introduction of a new production technology raises costs, other countries are pressed to make the same change on the grounds that the competitive balance between the introducer, say Germany, and other Member States must be restored.

The first amendment of the Treaty of Rome, the Single European Act (SEA) of 1986, was the vehicle for a number of items additional to, though not entirely independent of, the moves towards economic and monetary union for which the Act and its successor, the Maastricht accord, are so notorious. Among the items covered in the SEA was environmental policy which now covers air, land, sea, inland waters, fauna and flora. One influential result for forestry was the introduction of environmental assessment, and the consequential British law on the matter appeared as the Environmental Assessment (Afforestation) Regulations 1988. Some flexibility in environmental matters is clearly necessary and feasible where intra-Community competition is not affected: thus the Single European Act allowed that protective measures agreed for all should not prevent a Member State from having more stringent measures. An example is the protection of bird species where the number of protected species in the EC is greatly exceeded by the number on the British list under the terms of the Wildlife and Countryside Act 1981.

Britain is not alone in this involvement of environmental groups with forest policy and practice and in some other countries the pressures have been both greater and arguably have had even more effect. Thus in Germany, there is strong support from the lay public, as well as many biologists and some foresters to the movement in favour of a more natural forest (see, e.g. Otto, 1990). In the United States, the changes affecting management in the Federal Forest Service have been dramatic. They range from the identification of Roadless Areas in which logging is banned, to the judicial decision that the Service could not fell stands unless they could be shown to be declining in vigour, and most recently to the banning of timber sales in Federal Forests of Oregon and Washington where the northern spotted owl habitat is to be protected, with the result that the cut planned for these two States in 1991 was reduced by 32% (Rockwell, 1990). This represents a much more extreme position than that which would result from the attitude reflected by Smyth (1990) who notes that foresters 'face a tremendous challenge to prove to that (sceptical) public that management can mean working with nature, not against nature.'

The exploitation, conversion and sale to developed countries of tropical timbers is a major issue which has reached the political agenda. Its repercussions on forestry in Britain have thus far proved to be insignificant. While moist tropical forests currently rank in laymen's minds as the main concern for conservation of biodiversity, the movement is also strong in relation to the wet temperate forests of North Western America. The artificiality of almost all forest types in Europe means that the defence of wholly natural or supposedly natural forest is limited to a small proportion of total area. The treatment of 'ancient woodland', and particularly that with seminatural tree cover, is controlled in Britain. However, Britain's adherence to the Biodiversity Convention agreed at the Rio conference of UNCED in 1992 may have implications for the extension of forest on to certain land types which could be awarded enhanced conservation status.

An active afforestation policy with mainly timber-producing aims clearly widens the range of potential conflict with wildlife interests. Judging how scientific interest in particular habitats may alter is a testing business but there can be no doubt about the necessity for forecasting activity in this field if damaging conflicts with other land use interests are to be avoided or moderated.

The sensitivities noted apply equally to the conservation of archaeological monuments. As with wildlife conservation, archaeology may be liable to changing perceptions about the relative values of sites from different periods.

Recreation

The principal issue in woodland recreation which is of interest in Britain is that of wider access and the use of woodland for informal recreation. Because private landowners are generally unwilling or unable to offer free access, the burden of satisfying a claimant demand for access has fallen on public lands. Johnson and Nicholls (1991) record that one-third of estate owners in their sample survey in England and Wales provided recreation facilities: no mention is made of general public access. Some revenue has been generated by activities stimulated or made possible by the increase in this sort of traffic. Changes will no doubt occur in the nature of the experiences which countryside visitors wish to enjoy and market research and demand analysis can give useful pointers by assessing growth in different kinds of activity and the revenue-generating possibilities.

The issue of the degree of responsibility that lies with occupiers of private woodland is little discussed in Britain. Shoard (1980 and, with more vehemence, 1987) represents one of the more radical of critics of current practice, advocating legislation which would confer 'the right to roam' (a phrase Shoard borrows from H.G. Wells), including the right of access to all woodland. Hummel (1992) has suggested as one route to increased access that payment of Forestry Commission grants could be made contingent on such provision. This is, however, an improbable change in the current political climate and may be questioned on the practical ground that the great need for increased provision occurs mainly in the lowlands where grant applications arise in an almost random manner and at a low frequency. Property rights and the values that they sustain are crucial in this matter and it is certain that many owners in counties where free public access on foot is tolerated by convention would regret the limitation that such indirect pressure would imply. The Countryside Commission's (1987) proposals take the form of offering an alternative by way of community woodlands (see also Forestry Commission, 1991b) and the Commission subsequently favoured the notion of a Midlands Forest (Countryside Commission, 1990) aimed principally at enhancing the landscape of a depleted area of England but also requiring that the woodlands created should be accessible.

Employment, rural development and land use change

The provision of extra jobs and income has often been regarded as one of the most important values of forestry in place of extensive pastoral agriculture in remote areas. Despite the small gains in these measures from forests (as calculated by Johnson and Price, 1987 for Snowdonia and in a more limited manner by Mather and Murray, 1987 for private forestry establishment in Scotland) and the general dismissal of such considerations in forestry discussions of recent years (cf. Gallacher, 1987; Agriculture Committee, 1990), such influences are likely to rise in importance once more owing to changes in agricultural support. The costs and labour requirements of afforestation of lowland farmland are likely to be substantially higher than those of extensive upland work. Unfortunately, some of the argumentation in favour of afforestation on employment grounds (see, for an expression of the need for a single Community policy for rural space and the potential role of forestry, Fennell, 1988) is as tendentious as that against it from conservationist circles. A critical issue in this regard is whether jobs in one region must necessarily be at the expense of jobs elsewhere or not (Pearce, 1991), a point on which continuing debate is likely.

A second, quite separate matter concerns community integrity, part of the benefits of forestry investment which Fennell (1988) mentions. This is a perfectly legitimate benefit to claim, as legitimate as, say, the preservation of landscape. Unfortunately no one has assessed the community's willingness to pay for it in other than political terms.

Thirdly, agricultural adjustment to the changes in prices and production levels in prospect as a result of shifts in the Common Agricultural Policy constitutes a new challenge to forestry. Many forecasts of newly available land for non-farming uses have been founded on the naive calculation: surplus production divided by average yield equals land area surplus to farming. The issue here concerns the alternative purposes to which land may be put and the opportunity provided for tree planting.

Farmers are the other group beside environmentalists who express concern over the expansion of forest cover. Publicly, the agricultural industry may resent the loss of prestige and command over resources implied by a continuing shift of resources, especially land but also rural labour and government funds, into forestry. Forests are alien to most British farmers. This long-standing attitude originates from the time when the bulk of farms were tenanted. The tenant had, and usually has, no proprietary interest in timber which was instead the concern of the landlord. After the First World War, the major swing from tenanted to owner-occupied farms meant that farmers were unequipped by knowledge, experience and habit to carry out woodland management. This position remains broadly unchanged 80 years later. But against the loss of the land resource, weight must be given to the benefit enjoyed by the support provided to land prices, always a very significant element of farmers' wealth. The extraordinarily slow development of a policy of transferring resources from agriculture to forestry on better land reflects the political difficulties in adjusting the Common Agricultural Policy of the European Community. These are real, if peculiarly selfish with regard to repercussions on Third World countries, and on present showing afforestation is unlikely to have a significant impact, especially in view of the small incentive as measured by the low real values of wood price obtaining in recent decades.

Water

There are two issues in relation to water and the influence of forests in Britain and in both there are marked differences between the situations in upland and lowland areas. The first, quantity, continues to be the subject of intensive research. In the uplands, work at Balquhidder in Perthshire indicates that the differences in losses from evapotranspiration in forests compared with those from the seminatural vegetation are substantial and yields were reduced by similar proportions to those found in earlier work on the Welsh (Plynlimon) catchments (Calder and Hall, 1992; Johnson, 1992). In the lowlands, work in Hampshire shows losses by evapotranspiration from beech woods to be smaller than from intensively managed grass (Neal *et al.*, 1991).

The second concern is with water quality. In the uplands concern arises from the capture of acid inputs by tree crowns, with the greatest impact occurring in cloud (Forestry Commission, 1991c). The effect is local and significant on hard, base-poor rocks where the impact on insect larvae and young fish is marked (Department of the Environment, 1991, 1992, Hornung and Adamson, 1991). But it is clear that the objection to plantations in such areas does not stem solely from anglers; there are also those who have a deep conservation objection to the transfer of more heathland and moorland to plantations. 'Critical loads' of acidic pollutants are judged to be smaller for soils than for water in the affected areas. It is here relevant to note the Nordic experience where fish deaths have occurred because of higher acid inputs since the 1970s; in these countries impacts on tree and other plant growth are unclear but are believed to be trivial. In the lowlands, woodlands offer a means of diluting the effects of excessive nitrate application by farmers, an especially important consideration in view of the European Communities legislation on the reduction of nitrate additions to groundwater.

Wood production

Carbon storage

A new issue, unforeseen by foresters until the mid-1980s, and only of passing interest to researchers before that, is global warming. This may influence the growth of forests, and the forest and the use made of it are important in terms of carbon storage and hence mitigation of its effects (Grayson, 1988, Thompson and Matthews, 1989). This is the most dramatic illustration of developments with potentially large effects on policy and practice of any that have emerged for many decades. It provides an impressive example of the unpredictability of factors affecting policy. In the cases of air pollution and its supposed effects on trees, and of stratospheric ozone depletion and predicted effects on human health, politicians have found themselves under heavy pressure to act. In these two fields, their responses have been relatively rapid. This is especially true of ozone depletion where the costs of phasing out chlorofluorocarbons are not particularly high. In both fields, the decisions taken have led by several years, if not decades, scientific certainties of the actual change which is threatened. They are also ahead of the assessment of possible palliatives or controlling measures. Whereas the air pollution factor is potentially influential because it is pervasive and its effect, whether putative or proved, can, in some countries at least, be seen in sickly looking trees, that of impending climate change associated with warming is predicted, not actual. And confusingly, its effects may be beneficial in some regions, but because the prospective changes on sea level, water supply and above all crops are so widespread the issue is the more worrying to leaders and peoples in many different countries.

A whole raft of physical and biological research enquiries has been floated. But one vital concern, that of clearly stating the possible role of forestry, requires almost no research. Too much uncritical and misleading assertion is unhelpful, and may well be seriously damaging in any field of endeavour. Forestry stands in that uncomfortable position today. In particular, naive claims have been made about the value of extra tree planting for carbon storage which have not considered the global impact on management of the growing stock through effects on wood price.

Wood for energy

One of the few new silvicultural systems considered in recent years has been that designed to produce wood for energy using short rotations of coppice, often grown on relatively fertile land. Old-style coppice management, with its attractions to the small owner who could work it for a variety of products with the minimum of tools, has come full circle with the designs of Sirén *et al.* (1984) and others (see for example Hummel *et al.*, 1988). The case for major promotion of these recipes among the landowners with the favoured types of land depends crucially on a rise in the price of energy. Market forces have done little to aid the enthusiasms of supporters of the concept. Even branchwood, a cheaper material to supply, has not found a ready market and in any event price would often lead to the sale of such material to the particle board or pulp industries. Although a major upswing in fuel price is possible, for example as a result of a flight from nuclear power, this is unlikely to occur rapidly and the response time for short rotation coppice is reasonably short by the standards of the power supply industry. A more likely source of support for this form of forestry lies in the direction of a credit on carbon storage on the grounds that this is desirable as a palliative in the face of global climate change (Anderson, 1991). Barrett (1991) notes that Finland, The Netherlands and Sweden have introduced carbon taxes: the further step of introducing an explicit carbon credit remains to be taken. As with changing recreational demands, and even some conservation demands, the prospect for making timeous adjustment of forestry practice in all types of woodland in the light of policy shifts in this field appears reasonably good. These changes could involve species, thinning and felling age.

Wood-using industry

The interests of wood-using industry are normally expressed in unashamedly commercial terms. Industry has shown itself capable of adjusting to new specifications of wood in the medium term. The principal need is for stability – of volume and quality of wood supply and of price. Most firms would prefer to see an expanding supply without many new businesses entering the market. Given that competition policy will be an increasingly active component of the European scene, the prospects of stability in the sense of all existing enterprises continuing to trade must diminish to some extent. This will often tend to keep product prices down but should be favourable for roundwood prices to the extent that the more efficient operators survive. It is noteworthy that uncritical claims about an impending wood shortage are not heard as frequently as in the past. The weight of evidence at least for some decades into the future is not in favour of this as a cause of major price rises (Arnold, 1991 provides an authoritative assessment on this point). One argument which is promoted as might be expected is the autarchic one. A stark expression of this has been put by Lyden (1989) who notes Britain's high import bill for wood products (almost £7 billion in 1989) and states that: 'we have in our hands the potential, albeit in the slightly longer term of displacing at least 50% of that in value. . . . In this country . . . we have 2 million hectares of farming land surplus to requirement. What better purpose could that land be put to than forestry.' The objective of import substitution *per se* is indefensible to economists (see, for example, Pearce, 1991) and is not supported by government officials, but tends to be strongly supported by politicians of all hues in Britain.

The financial power wielded by wood processors is so great that the influence of this pressure group is substantial; it is even more relevant in regions such as Canada and north Sweden where the industry is the largest provider of earnings in isolated regions. Although such a dominant position is uncommon in western and central Europe, the industry's countervailing strength in relation to the claims of competing outputs of the forest remains high and this is especially influential in countries with established forest where the conservation argument centres on the handling of the existing growing stock. The long interval between planting and wood supply means that industrial arguments have less impact on the location of afforestation except in broad terms but they are important in relation to the proposed overall scale of planting. This is a token of the confidence of government and private investors in the future, a token which wood-processing firms are eager to recognize.

The Exchequer

Forestry and, in particular, afforestation have suffered more than most industries from doubts about the validity of government expenditure on their support. Before more critical investment criteria were introduced by the Treasury, assessments of state afforestation were allowed to continue broadly unchecked on the basis of claims concerning additional employment opportunities, the desirability of maintaining rural income and services, and vague expectations of recreation activities and the growth of regional wood-using industry. In 1970, following a change of government, Mr (now Sir Ted) Heath decided that a revision of the financial reporting system of the Forestry Commission should await an assessment of the virtues of continued state and private afforestation on the basis of a comprehensive cost–benefit appraisal. This was duly carried out (HM Treasury, 1972). In the light of the less than clear-cut outcome and of the considerations of an interdepartmental committee formulated with a view to restricting the growth of government commitment, Ministers decided to allow planting to continue but on a declining area target (Anon., 1972; Forestry Commission, 1973). Subsequent administrations conducted enquiries on the return to public investment in new planting. The most searching of these was that conducted for the National Audit Commission by Planning, Economic and Development Consultants (1986) following enquiries by the powerful Committee of Public Accounts in 1986–1987. The government, responding to the Committee (Forestry Commission, 1989d), accepted that more effort should be devoted to assessing the value of the non-market benefits produced. Neither report was decisive in influencing Ministers' attitudes towards new planting except in the anxiety to give greater emphasis than ever before to ensuring that the effects of afforestation on the environment were positive and to restate the desirability of continued afforestation, as illustrated by the confidence-building announcement by the Secretary of State for Scotland (Forestry Commission, 1989c) following the 1988 announcement of the exemption of commercial forestry from income taxation.

Successive government reviews both of the value of afforestation and its more efficient implementation have been carried out against the background of a continual rumbling from dissenting voices. (A comprehensive survey of policy proposals for forestry which emerged in the 1980s is given by Gilg, 1992.) The interest of the Nature Conservancy Council has already been mentioned. Most commentaries have not considered in detail, or made fresh, economic analyses, instead relying on the selection of elements in favour of their chosen position. These have come from political bodies such as the Institute of Economic Affairs (Miller, 1981), from industry bodies or sympathizers (FICGB, 1990, Denne *et al.*, 1986), from environmental bodies and notably the Ramblers Association (1980), the Council for the Protection of Rural England (PIEDA, 1987), and a variety of commentators from academia (e.g. Stewart, 1988). However powerful the arguments appear for curtailing government expenditure on forest expansion, policy remains strongly positive on this front.

A distinctive feature of policy reviews over the past 20 years has been the attention given to valuing the non-market benefits of forestry, with particular reference to new planting; little work has been done on the implications for management of existing forests.

Forestry industry

The response to new demands

In the face of so many demands which influence and usually detract from the achievement of the traditional goals of growing wood for gain, the forestry industry itself has developed in a number of significant ways. One has been to capitalize on the increased demand for access and recreation and to win friends by opening woodlands. Responses have differed between public and private sectors, with the more controllable and revenue-producing forms favoured in the latter. Secondly, a more pro-active attitude to the claims of outsiders on the desirable form of silviculture and management has been adopted. Some critics of existing practice would go much further by advocating a major expansion of continuous cover silvicultural systems (Helliwell, 1991), but the more extreme 'near to nature' form of forestry promoted in continental Europe has not attracted the same support in Britain. Thirdly, forest interests have a further opportunity in developing new concepts of forests' contribution to the landscape and urban society through, for example, the creation of new types and styles of woodland, as illustrated by, for example, the concept of community woodlands as outlined in Forestry Commission (1991b). In practice, there has been significant convergence of the position of the environmentalists noted at the beginning of this chapter and that of foresters. Nevertheless there remains the major issue of the essential arbitrariness of the extent of adjustment of management from the course which all but a philanthropist would otherwise adopt.

The balance between private and other forestry

An additional issue concerns the ownership of forest land, a topic which has become a particularly active matter in Britain since the passage of the

Forestry Act 1981 which empowered the Forestry Commission to dispose, subject to certain exceptions, of any of its plantations. This study makes no analysis of the desirable balance of private, state and local authority woodland, but recognizes that decisions on this issue are crucial for goal setting and implementation.

A feature of private forest holdings is the increasing extent of ownership by charitable trusts, societies and the like. It should also be remembered that government departments apart from the state forest enterprise, together with nationalized industries, local authorities and other public sector bodies own woodland. For many woodland owners or intending owners the existence of alternative buyers and sellers is a sign of a maturing industry, increasingly offering opportunities for profitable exchange.

The question that is germane to the present enquiry concerns the role of private woodland owners in relation to whatever national policy is decided by government. There is, at least in principle, no difficulty over those demands that are mediated through the market: an owner may decide to produce the commodity in demand or he may not. A fundamental issue so far as owners are concerned is that of balancing the incentives to produce marketable products, principally but by no means exclusively wood, with the provision of those non-market benefits whose supply cannot reasonably be considered part of the conventional responsibility of landownership. Property rights have become much restricted over the years with recent administrations adding to rather than subtracting from the rules by which landowners manage their estates. At the same time certain compensations have been paid. Thus Nature Conservancy accounts show an increase in annual and capital grants for management of all sites of special scientific interest, etc. from £1.1m in 1980 to £9.2m in 1989. An essential question that remains is how far is it possible and appropriate to engage private landowners in further adjustment of management. Discussion of this central question must be postponed until the final chapters.

Government Policy on Private Forestry

The foregoing listing of issues reflects the principal themes actively discussed during the 1980s. Not all of them find expression in the current policy of government although they may all underlie government's determination to continue with expansion of forests and woodlands and the diversification of the form of those new woodlands and of existing woods, combined throughout with a favouring of increasing private participation in all aspects of the forestry industry.

There is no separate statement of government policy for private forestry in Great Britain. In 1991, the government published (Forestry Commission, 1991d) a document setting out a 'composite summary' of its current forestry policy. An important concern is with the need for government to 'carry' constituents of the diverse groups which influence government ministers and the forest authority in its day-to-day dealings with them.

The current statement contains eight sections, the first of which is the most general, and each of which partly depends on photographs and captions to convey the message of a concerned policy with a markedly environmentally friendly emphasis. The following summary selects those features which relate particularly to private forest owners' concerns.

1. Aims – 'The two main aims of the government's forestry policy are thus: The sustainable management of our existing woods and forests. A steady expansion of tree cover to increase the many, diverse benefits that forests provide'.
2. The benefits of positive management of woodlands:
   (a) encourage management of existing woodlands to yield a wide range of benefits including production of quality timber;
   (b) bring long-neglected woodlands under management;
   (c) re-structure forests (that is, adjust the lay-out of restocked stands to meet landscape and other objectives);
   (d) encourage opening of woods to visitors;
   (e) arrange for continuing access to woodland sold by the Forestry Commission;
   (f) maintain general presumption against conversion of all kinds of woodland to agriculture by strict application of felling licensing.
3. Expanding the forest area:
   (a) there is scope for new planting to take place at the indicative level of 33,000 ha a year (for State and private combined) announced in 1987;
   (b) support via a flexible scheme tailored to multiple purpose objectives.
4. Preferred areas for new planting:[1]
   (a) select land where planting will be an environmental gain;
   (b) maintain the presumption against large-scale planting of predominantly coniferous species in the English uplands, but resist extension of this presumption to Scotland and Wales;
   (c) use agreed 'indicative forestry strategies' in Scotland to decide where planting may proceed, and investigate potential for extending the indicative strategy system to England and Wales;
   (d) encourage new planting in lowlands and close to centres of population to create 'Community Forests'.

[1] It will be noted that this section is given special emphasis by separating it from the previous one.

5. Taking account of environmental needs:
   (a) maintain broadleaves policy[1];
   (b) maintain and improve procedures for consultation with interested parties, especially those concerned with amenity and conservation;
   (c) continue research on effects of air pollution working through tree canopies to cause water acidification.

6. Streamline the Forestry Commission:
   (a) reorganize to reflect different objectives of the Forestry Authority and the Forest Enterprise.

7. Forestry overseas:
   (a) participate in a constructive way in European Community discussions on the development of international forestry instruments.

8. Delivering public benefits:
   (a) remain firm in support of British forestry;
   (b) increase emphasis on social and environmental benefits and tailor support to delivery of benefits for the community at large.

The word 'environment' occurs frequently in the statement and environmentally attractive scenes form the vast bulk of the document's illustrations. Wood production and the generation of income through market activities of owners are conspicuous by their absence. This balance may be questioned for its lack of realism.

Administration

In April 1992 the Forestry Commission's functions were clearly separated through the establishment of the Forestry Authority, concerned with regulation and advice, staffed by a different group of officers from those forming the Forest Enterprise and responsible solely for management of state forests (Lang, 1991). Before this, managers of the rank of District Officer and above had combined the two functions although these had long been identified and separate accounts had been prepared for the two since 1966. The Authority has a Commissioner responsible to the full Forestry Commission board, three (country) Chief Conservators and 27 Conservators. These officials are responsible for the systems of control of felling in all woodland including Enterprise forests, the administration of grant aid and associated advice. As before, the Research Division remains within the Authority.

---

[1] Which requires that the area of broadleaved forest in the country should not diminish and that any broadleaved area felled should normally be replaced by broadleaves.

Three 'countryside' agencies operate in Great Britain, namely the Countryside Commission, the Countryside Council for Wales and the Scottish Natural Heritage agency. These bodies offer grants, in combination with local authorities, for planting of trees and woodlands of under 0.25 ha.

The representative body for private woodland owners is Timber Growers UK (TGUK) which also covers Northern Ireland. Although its membership currently accounts for only 30% of the area of non-Forest Enterprise woodland, its members include the bulk of the owners of large estates and a wide range of other owners, including local authorities.

WOODLAND OWNERSHIP IN BRITAIN

The policy statement makes no direct reference to the different clients for government support in the 'private' sector. This group of existing (or potential) owners in fact covers all individuals, partnerships, trusts, firms and agencies other than the Forestry Commission. In practice it is believed that woodland holdings of central government are small and will decrease. However, those of local authorities are predicated to expand, alongside the larger extent of strictly non-governmental ownerships.

It is clear that there must be a very wide range of objectives among woodland owners. It is a reflection of the lack of attention to the subject of ownership, its composition, geographical and size distribution, etc. that no comprehensive data have been compiled even on such basic statistics as the numbers and extents of woodland holdings in Britain. There is thus a general lack of statistics enabling any assessment to be made retrospectively on the effectiveness of past programmes of government support. The position is in process of change since data are now to be compiled on the ownership category of those joining the current Woodland Grant Scheme. This is a welcome innovation if any useful policy analysis, an activity given emphasis in this work, is to be undertaken.

Data on the kinds of owners of agricultural land, which may include woodland, have been compiled from time to time but these offer no useful guide to either the woodland component or its relationship to the rest of landowners' property. Locke (1962) made a bold attempt to assess the likely distribution of woodland ownerships by number and area from the revision of the 1947–1949 census of woodlands undertaken in 20 counties between 1952 and 1959. The results must be regarded as extremely tentative but are notable in providing at least one benchmark measure. The data given in Table 5.3 refer to a year before the major burst of post-war afforestation by private investors.

The size structure of private woodlands covered by plans of operations agreed with the Forestry Commission is known from the mid-1970s, and is set out in Table 5.4. The distributions shown do not necessarily refer

Table 5.3. Size distribution of private woodland holdings in 1962, given as a percentage of the total area.

	Size class (ha)					
	<10	10–20	20–40	40–200	200–400	400+
Eng. and Wales	12	9	13	33	14	19
Scotland	1	3	4	27	20	45

Source: Locke (1962).

Table 5.4. Size distribution of dedicated and approved woodlands estates, given as a percentage of the total area.

<table>
<tr><td></td><td colspan="5">Size class (ha)</td></tr>
<tr><td></td><td><40</td><td>40–200</td><td>200–700</td><td>700–2000</td><td>2000+</td></tr>
<tr><td>Eng. and Wales</td><td>10.7</td><td>40.4</td><td>28.8</td><td colspan="2">20.1</td></tr>
<tr><td>Scotland</td><td>2.8</td><td>22.8</td><td>34.6</td><td>28.1</td><td>11.7</td></tr>
</table>

Source: Economic surveys of private forestry (Aberdeen University, 1973; Commonwealth Forestry Institute, 1973).

to separate management units. This is because the data for plans are based on units which ranked at the time as separate plans but might have been managed in association with other nearby properties of the same owner or family. The total areas concerned were 262,000 ha in England and Wales and 210,000 ha in Scotland.

No more recent data on the area of Forestry Commission aided woodlands are available for such a comprehensive set. This is largely because the nature of the woodland grant schemes has altered and become partial instead of more comprehensive in the coverage of a given property's woodlands.

Membership statistics from Timber Growers United Kingdom show no separation by ownership type but provide evidence on the size structure of separate estates' woodlands (Table 5.5). Total areas for Britain were 384,000 ha in 1982 and 408,000 ha in 1991.

Data are not available for whole woodland estates within Forestry Commission schemes but records are compiled for current planting schemes by ownership category. Table 5.6 indicates the distribution of a recent cross-section of owners entering the Woodland Grant Scheme. Although the original data show 16 categories of owner and 'other', the figures, which refer to a total area of 88,000 ha of areas accepted into the Scheme, have been grouped for convenience.

Table 5.5. Size distribution reported for timber growers' members estates, given as a percentage of the total area.

	Size class (ha)					
	10	10–20	20–40	40–200	200–400	400+
	Percentage of total area					
1982 data						
England and Wales	neg.	1	1	32	21	42
Scotland	neg.	1	3	22	21	53
Great Britain	neg.	1	3	27	21	47
1991 Data						
GB	0.3	1.0	2.6	26.5	18.2	51.4
	Percentage of total numbers					
GB	7.0	10.6	14.6	47.5	18.2	8.5

Sources: 1982 Annual Reports (Timber Growers England and Wales, 1982; Timber Growers Scotland, 1982); Timber Growers UK (personal communication).

Table 5.6. Distribution of recent Woodland Grant Schemes by owner.

Category	Number (%)	Area (%)
Individual, family partner, executor	65	57
Trust, syndicate	22	30
Government dept., Local authority	6	2
Charity, National Trust	2	4
Crown Estate, Church Commissioners, Royal Estates	0.5	0.7
Other	4	6

Source: Forestry Commission (personal communication).

The indications from this evidence are that there is no very large difference in the area per application for entry to the latest scheme (the Woodland Grant Scheme). The bulk of the activity is generated by individuals, families, executors and trusts (which often relate to family combinations). For the total population of woodland estates there is clearly a very wide range in size of holding at least as measured by area, but nothing can be inferred about the distribution of areas by type of owner.

It should be remembered that the statistics in all the above tables include the properties of all but the Forestry Commission, that is they include those of other central government departments, local government, public utilities, nationalized corporations, and the Crown Estates.

No analyses are possible of the distribution of payments among the different categories and sizes of holdings or of other responses to government regulations.

Implementation of Policy

Until 1988, when intervention at the very highest level of government led to the abrupt decision to alter the tax treatment of investments in forestry operations, a mixed regime of tax offsets and (taxable) grants was in place to support private forestry. This had been highly successful in achieving a high rate of investment in new plantations, but at the cost of a loss of public support owing to concern over the scale of individual schemes and particularly the size of the fortunes involved. The high rate of change in the appearance of the countryside, together with nature conservation worries, appear to be the origin of the heavy emphasis on the adoption of environmentally attractive ways of creating new woodlands in the policy outlined above. Following the 1988 decision (Forestry Commission, 1989c), support was confined to grants alone, with the favourable tax treatment of woodlands being phased out by 1993, and the only significant direct tax remaining until 1992 being inheritance tax. These supports are set out below.

A significant step in the implementation of national government policy on new planting has been the agreement with local authorities of so-called indicative forestry strategies. The leading role in developing the concept and practice has been Strathclyde Regional Council (1992). The purpose is to state a strategy which integrates the interests of forestry investors with those of the local and wider communities. Such strategies usually apply to the broad location of afforestation but in principle any forest or forest-based activity may be brought within the content of a strategy. In 1992, the Department of the Environment in England and the Welsh Office consulted with English and Welsh counties and National Park Authorities on the preparation of similar strategies in their respective countries. A further development has been the issue of ‘Guidelines’ (e.g. Forestry Commission, 1991c) setting out desired methods of conducting operations. Approval of grant applications is conditional on the applicant agreeing to abide by these guidelines: this form of administrative action thus effectively has the same force as a legal measure.

The following paragraphs set out the position on forestry taxation and on grants payable under Forestry Commission schemes or schemes jointly arranged by agriculture Departments and the Forestry Commission. The details noted are correct as of 1 April 1992 in the case of grants and make allowance for proposed changes to be made in the Finance Act 1992 to amend the Inheritance Tax rules for businesses.

TAX

(see also Forestry Commission, 1989f)

Income tax

Commercial woodlands are not subject to income or corporation tax. Transitional arrangements apply to woodlands occupied under Schedule D of income tax at 15 March 1988 which applied until 5 April 1993.

Inheritance tax

Deferral of tax is available on timber but not the land transferred. The value of the timber is then left out of account in determining the value of the estate. When timber is disposed of subsequently, whether by sale or otherwise, tax is borne at the appropriate marginal rate. The scale of tax applicable is that in force at the date of disposal. If the beneficiary dies before all the timber existing at the time of the transfer has been disposed of, the remaining liability to deferred tax charge is wiped out.

Business relief is available for woodlands which have been owned by the transferor for at least two years immediately before the transfer. The higher rate of relief was originally set at 50% but was increased to 100% in 1992. This rate is applicable to a solely owned or to a life tenant's business, an interest in a business such as a share in a partnership, or unquoted shareholdings of over 25%. The lower rate of 50% (previously 30%) applies to land, buildings or machinery owned by the transferor or used in his business. Business relief at the rates noted above is available on deferred disposals.

Summary of tax position

The arrangements applying to commercial forestry since early 1992 effectively exempt woodland owners from income, capital gains and inheritance taxes. These circumstances provide a clear stimulus to potential purchasers of existing woodland seeking to reduce capital gains or inheritance tax liabilities. While the provision of a secondary market is beneficial to those investing in planting, the new arrangements must encourage the manipulation of woodland holdings solely for tax reduction.

GRANTS

The principal source of grant aid is the Forestry Commission.

Table 5.7. Woodland Grant Scheme.

	£ ha^{-1}	
Size of planting area (ha)	Conifers	Broadleaves[a]
less than 1.0	1005	1575
1.0–2.9	880	1375
3.0–9.9	795	1175
10.0 and over	615	975
Better Land Supplement[b]	400	600
Community Woodland Supplement[c]	950	950

[a] Planting of native pinewoods ranks for the broadleaved grant.
[b] Available for new planting of arable land and improved grassland.
[c] Available under conditions designed to ensure that the new planting undertaken will enhance recreational opportunities close to towns.
Sources: Forestry Commission, 1991c, 1992.

New planting and restocking under the Woodland Grant Scheme

Plantations rank as broadleaved where the proportion of such species is over 50%. The relationship between grant and size of planting is given in Table 5.7.

Grants are payable in three instalments for planting, namely 70% after planting, 20% 5 years later, and 10% 10 years after planting. For natural regeneration, the proportions are 50% on completion of the approved work, 30% after adequate stocking is achieved, and 10% 5 years later.

Management

Standard management grant is paid for agreed operations during the normal maintenance period following the initial establishment phase. Special management grant is payable for operations in woodlands of special environmental value from not earlier than 11 years. Rates are as shown in Table 5.8.

Management plan

A grant of £100 is payable for preparation of a plan for woodlands eligible for management grants and not previously grant-aided by the Forestry Commission.

Table 5.8. Management grant.

	£ ha^{-1} year^{-1}		
	Standard		
Size of wood (ha)	Conifer	Broadleaved	Special
Less than 10	15	35	45
10 and over	10	25	35

Table 5.9. Farm Woodland Premium Scheme.

	£ ha^{-1} year^{-1}		
		Less-favoured areas	
Type of land	Lowlands	Disadvantaged	Severely disadvantaged
Arable or improved	250	190	130
Unimproved grassland	not available	60	60

New planting under the Farm Woodland Premium Scheme

Grants are as for the Woodland Grant Scheme except that minimum application size is 1 ha although there is no minimum individual block size, and may include Better Land Supplement where appropriate. In addition the annual payments shown in Table 5.9 are made for 15 years for woodlands containing more than 50% by area of broadleaves and 10 years for woodlands containing 50% or less of broadleaved species.

Setaside

Land accepted by agriculture Departments as Setaside under the scheme available from 1988 is not eligible for entry into the Farm Woodland Premium Scheme; instead planting is grant-aided under the Woodland Grant Scheme and the appropriate Setaside annual payments are then payable.

Countryside agencies' schemes

The three countryside agencies offer assistance for tree and small woodland planting on areas under 0.25 ha. Typically the aid amounts to 40% of the cost of plants and planting, with specific amounts paid for fencing. The cost of government support of such schemes is shared with local authorities so that the aggregate of public expenditure on them rises to 60–70% of the cost.

Total expenditure

Forestry Commission expenditure on grant schemes including administration was £18.8 m in 1989–1990 and £21.9 m in 1990–1991. Comparable data are not available for other agencies' expenditure but this is believed to amount to not less than £5m annually.

EXTENSION

The Forestry Commission and the countryside agencies offer guidance on the grant schemes they administer, together with broad advice on appropriate practices. Agriculture departments also undertake some extension work as a result of the recent initiatives on planting of farm woodlands. No data are available for total expenditure on extension services. Excluding purely administrative aspects, such as consultation with interested parties on grant schemes but including provision for part of the information activities of government bodies, the cost of extension activities is estimated to be not less than £3m per year.

RESEARCH

The Forestry Research Coordination Committee (1992) provides an annual statement of expenditure on research by all public agencies known to finance forestry research in Britain. Excluding the cost of work on tropical forestry, this amounted to £20.55m in 1991; £9.65m of the total was financed by the Forestry Commission. In years to 1990–1991, 50% of the expenditure on Forestry Commission research was allocated to state forests and 50% to private on the basis of assessment of the balance across each field of research. (The change to a 32%/68% split in accounts for 1990–1991 (Forestry Commission, 1991f) may be assumed to reflect the desire to reduce charges on the accounts of the forest enterprise and to indicate that a larger proportion of the whole could only be ascribed to the needs of the industry as a whole.) Almost all the research supported by agriculture and environment departments, totalling £6.1m, relates to the needs of farmers and local authorities respectively. Assuming 90% of their expenditure is oriented to the needs of non-state forestry, and that other agencies' expenditure relates equally to state forest and other, gives a total allocable to private forestry of £12.7m.

TOTAL PUBLIC SUPPORT OF PRIVATE FORESTRY

The figures below for expenditures arising on account of public financing of support to private tree planting activity and forestry are estimates. Figures are given in £m $year^{-1}$ (1991).

- Grants 27
- Extension 3+
- Research 13
- Total 43

COMMENTARY ON THE IMPLEMENTATION OF POLICY

There are few published studies on the effects of government support on the scale of private woodland management in Britain. Despite this lack, industry commentators have always been quick to suggest the reasons for fluctuations in the critical parameter of post-war policy, the annual rate of new planting.

A benchmark enquiry was that of Nicholls (1969) whose survey of 72 English and Welsh estates collected information on relevant facts on the estate including the woodland economy. The principal conclusion drawn from the correlations calculated was that the value of the tax concessions to forestry was identified as the most important determinant of woodland management.

The removal of forestry from income tax has clearly reduced interest in new planting by private owners. The rate fell from a post-war peak of 25,400 ha in 1988–1989 to 15,500 ha in 1990–1991 and to 14,200 ha in 1991–1992. The 1988 announcement of the tax change was the stimulus for a complete updating of a subsequent survey undertaken between 1985 and 1987 by Johnson and Nicholls (1991). The purpose of the study was to assess the influence of taxation and grants on the management of woodlands on the same group of estates as had been surveyed 20 years before, together with a sample of farms. Low rates of return on woodlands, after government support, discouraged investment. For estate owners the tax arrangements were the more important form of assistance, for farmers the opposite applied. Estate owners were increasingly prepared to undertake tree growing for purposes other than timber production.

A number of other commentaries have been made on the progress of Forestry Commission schemes. Many of these have been carried out by geographers, and although useful descriptive material has been compiled the effectiveness of the schemes has not been a central feature of the studies. Thus Mather and Murray (1988) reviewed the agents of afforestation, that is owners, management companies, consultants or others, and also considered employment effects. Watkins (1984) found from a survey in the period 1978–1981 that 72% of estate owners interviewed had never received a planting grant and, in line with much anecdotal evidence, that there was considerable ignorance of the grants available. Firn Crichton Roberts (1990) carried out a study for the Forestry Commission on a sample of 110 landowners or their advisers to assess the success of the Woodland Grant Scheme following the 1988 tax change. A principal finding was that the design of the grant scheme required greater understanding of landowners' objectives, and different forms of promotions of forestry investment needed to be shaped to suit the different potential investors.

Gasson and Hill (1990), in a survey of samples of participants and non-participating farmers in the Farm Woodland Scheme set up in 1988, found that the scheme's contribution to the reduction of agricultural output was modest, and that the consumption motive (shooting, etc.) took precedence in the design of plantations over the wood production motive. The (public) cost-effectiveness of the scheme appeared to be small though positive.

An important line of enquiry concerns the process of consultation following receipt of an application by the Forestry Commission from an owner wishing to fell or to plant a specific area. This topic became prominent because of complaints about the less than transparent procedures adopted before permission was granted for substantial changes to the landscape and the nature of the activity it was to absorb. Shoard (1987, Chapter 10) described the nature of the problem, and Brotherton and Devall (1988) and Mather (1989) illustrated the results of statistical enquiries into the numbers of schemes proposed, those subjected to review by Forestry Commission processes and those reviewed by Ministers. Such studies are sometimes difficult to interpret; for example a number of proposed planting schemes are never pursued to the point of application for grant aid, and a further proportion are varied in the course of the consultation process.

It is clear that there is a dearth of published evidence on the success, a concept which it increasingly difficult to define, of the various government schemes for woodland creation and management available to landowners in Britain. Current public expenditure on grants, extension services and research and development of more than £40m requires fuller assessments of the effectiveness of implementation measures, a matter further considered in Chapter 18.

References

Aberdeen University (1973) *Economic Survey of Forestry Costs*. Department of Forestry, Aberdeen.

Agriculture Committee (1990) *Land Use and Forestry*. Second report of the committee: Session 1989–90. House of Commons paper 16–1.

Anderson, D. (1991) *Carbon Emissions and Carbon Fixing from an Economic Perspective*. Forestry Commission and the Scottish Forestry Trust, Edinburgh.
Anon. (1972) *Forestry Policy*. (Statement of the forestry Ministers.) HMSO, London.
Arbuthnott, H. and Edwards, G. (1989) *A Common Man's Guide to the Common Market* 2nd edn. Macmillan, London.
Arnold, J.E.M. (1991) The long term global demand for and supply of wood. Paper No 3 in *Forestry Expansion: a Study of the Technical, Economic and Ecological Factors*. Forestry Commission, Edinburgh.
Barrett, S. (1991) Global warming: the economics of a carbon tax. In: Pearce, D. *et al.* (eds), *Blueprint 2: Greening the World Economy*. Earthscan Publications Ltd, London.
Brotherton, I. and Devall, N. (1988) Forestry conflicts in National Parks. *Journal of Environmental Management* 26, 229–238.
Calder, I.R. and Hall, R.L. (1992) *Growth and Water Use of Forest Plantations*. Wiley, London.
Commonwealth Forestry Institute (1973) *Survey of Income and Expenditure*. Department of Forestry, Oxford.
Countryside Commission (1987) *Policies for Enjoying the Countryside* (CCP 234) and *Priorities for Action* (CCP 235). Countryside Commission, Cheltenham.
Countryside Commission (1990) *A New National Forest in the Midlands* (CCP 278). Countryside Commission, Cheltenham.
Denne, T., Bown, M.J.D. and Abel, J.A. (1986) *Forestry: Britain's Growing Resource*. UK Centre for Economic and Environmental Development (CEED), London.
Department of the Environment (1991) *Acid rain – Critical and Target Loads Maps for the United Kingdom*. DoE, London.
Department of the Environment (1992) *Critical and Target Loads Maps for Fresh-Waters in Great Britain*. Department of the Environment, London.
Fennell, R. (1988) The evolution of rural policies in the European Community. Paper to Section K, British Association. *Oxford Forestry Institute Occasional Paper* no. 38.
Firn Crichton Roberts Ltd (1990) *The Attitude of Landowners to the Woodland Grant Scheme: a Report to the Forestry Commission*. Pittenweem, Fife, Scotland.
Forestry Commission (1973) Forestry policy. In: *53rd Annual Report of the Forestry Commission 1972–73*. HMSO, London.
Forestry Commission (1989a) Afforestation in the English uplands. Statement by the Secretary of State for the Environment. Appendix IX, *68th Annual Report of the Forestry Commission*. HMSO, London.
Forestry Commission (1989b) New planting in the uplands of England. Announcement by the Minister of Agriculture. Appendix VI, *69th Annual Report of the Forestry Commission*. HMSO, London.
Forestry Commission (1989c) Forestry policy. Announcement by the Secretary of State for Scotland. Appendix VII, *68th Annual Report of the Forestry Commission*. HMSO, London.
Forestry Commission (1989d) The Government's response to the report by the Committee of Public Accounts. Appendix XII *69th Annual Report of the Forestry Commission*. HMSO, London.

Forestry Commission (1989e) New support arrangements for private forestry In: *69th Annual Report of the Forestry Commission*. HMSO, London.

Forestry Commission (1989f) Taxation of woodlands. *Forestry Commission Bulletin 64*. HMSO, London.

Forestry Commission (1991a) *Forestry Facts and Figures*. Forestry Commission, Edinburgh.

Forestry Commission (1991b) Community forests, In: *70th Annual Report of the Forestry Commission*. HMSO, London.

Forestry Commission (1991c) *Forests and Water Guidelines*. Forestry Commission, Edinburgh.

Forestry Commission (1991d) *Forest Policy*. Forestry Commission, Edinburgh.

Forestry Commission (1991e) *Woodland Grant Scheme*. Forestry Commission, Edinburgh.

Forestry Commission (1991f) Forestry Authority Services. In: *71st Annual Report of the Forestry Commission*. HMSO, London.

Forestry Commission (1992) Press release 31 January 1992. Forestry Commission, Edinburgh.

Forestry Commission (Undated) *Census of Woodlands and Trees 1979–82: Great Britain*. Forestry Commission, Edinburgh.

Forestry Industry Committee of Great Britain (FICGB) (1990) *Options for British Forestry*. FICGB, London.

Forestry Research Coordination Committee (1992) *Collation of Forestry Research Programmes - 1992*. Forestry Commission, Alice Holt Lodge, Farnham.

Gallacher, Lord *et al.* (1987) EC forestry policy. *House of Lords Hansard* 484(41): cols 1033–1081.

Gasson, R. and Hill, G.P. (1990) An economic evaluation of the Farm Woodland Scheme. FBU Occasional Paper 17, Farm Business Unit, Ashford, Kent.

Gilg, A.W. (1992) *Countryside Planning Policies for the 1990s*. CAB International, Wallingford.

Grayson, A.J. (1988) Carbon dioxide, global warming and forestry. *Research Information Note* 146. Forestry Commission, Edinburgh.

Helliwell, R. (1991) Continuous cover forestry group. *Association of Professional Foresters*, December, 20–21.

HM Treasury (1972) *Forestry in Great Britain, an Inter-departmental Cost Benefit Study*. HMSO, London.

Hornung, M. and Adamson, J. (1991) The impacts on water quality and quantity. In: *Forestry Expansion: a Study of the Technical, Economic and Ecological Factors*, paper no. 9. Forestry Commission, Edinburgh.

Hummel, F.C. (1992) Aspects of forest recreation in Europe. *Forestry* 65, 237–51.

Hummel, F.C., Palz, W. and Grassi, G. (eds) (1988) *Biomass Forestry in Europe: a Strategy for the Future*. Elsevier Applied Science Publishers, Barking, Essex.

Inland Revenue. *Inheritance Tax*. (IHT 1.) HMSO, London.

Johnson, J.A. and Nicholls, D.C. (1991) The impact of government intervention on private forest management in England and Wales. *Occasional Paper* 30. Forestry Commission, Edinburgh.

Johnson, J.A. and Price, C. (1987) Afforestation, employment and rural depopulation in the Snowdonia National park. *Journal of Rural Studies* 3, 195–205.

Johnson, R.C. (ed.) (1992) Effects of upland afforestation on water resources: the Balquhidder experiment 1981-1991. *Institute of Hydrology Report* 116, NERC, Swindon.

Lang, I. (1991) Reorganisation of the Forestry Commission. Appendix IX, *71st Annual Report of the Forestry Commission*. HMSO, London.

Locke, G.M.L. (1962) Distribution of private woodland area by size of estate. *Quarterly Journal of Forestry* 56, 310–315.

Locke, G.M.L. (1971) *Forestry Commission Census of Woodlands 1965–67: a Report on Britain's Forest Resources*. HMSO, London.

Lowe, P. (1987) Environmental concern and rural conservation politics. In: Whitby, M. and Ollerenshaw, J. (eds), *Land Uses and the European Environment*, Belhaven Press, London.

Lyden, K. (1989) Towards a rethink of UK forest policy. In: *Forest Policy into the 1990s. Proceedings Discussion Meeting*. Institute of Chartered Foresters, Edinburgh, pp. 163–168.

Mather, A.S. (1989) *Afforestation Consultations in Scotland*. Report to the Forestry Commission, Department of Geography, University of Aberdeen.

Mather, A.S. and Murray, N.C. (1988) The dynamics of rural land use change; the case of private sector afforestation. *Land Use Policy* 5, 103–120.

Miller, H. (1991) British forestry in 1990. In: *Forestry expansion: a Study of Technical, Economic and Ecological Factors*, paper no. 1 Forestry Commission, Edinburgh.

Miller, R. (1981) *State Forests and the Axe*. Institute of Economic Affairs, London.

Moore, N.W. (1987) *The Bird of Time: the Science and Politics of Nature Conservation*. Cambridge University Press, Cambridge.

Nature Conservancy Council (1984) *Nature Conservation in Great Britain: Summary of Objectives and Strategy*. Nature Conservancy Council, Peterborough.

Nature Conservancy Council (1986) *Nature Conservation and Afforestation*. Nature Conservancy Council, Peterborough.

Neal, C., Robson, A.J., Hall, R., Ryland, G., Conway, T. And Neal, N. (1991) Hydrological impacts of hardwood plantation, Britain. *Journal of Hydrology* 127, 349–365.

Nicholls, D.C. (1969) Use of land within the proprietary land unit. *Forestry Commission Bulletin* 39. HMSO, London.

Otto, H.-J. (1990) [The ecological facts and objectives of a near-to-nature forestry.] *La Forêt Privée* 196, 37–43.

Pearce, D. (1991) Assessing the returns to the economy and to society from investments in forestry. In: *Forestry Expansion: a Study of Technical, Economic and Ecological Factors*, paper no 14, Forestry Commission, Edinburgh.

Pearce, D., Markandya, A. and Barbier, E.B. (1989) *Blueprint for a Green Economy*. Earthscan Publications, London.

PIEDA (1987) *Budgeting for Forestry*. Recommendations for reform of forestry taxation policy. Council for the Protection of Rural England, London.

Planning, Economic and Development Consultants (1986) *Forestry in Great Britain: an Economic Assessment for the National Audit Office*. HMSO, London.

Ramblers Association (1980) *The Case Against Afforestation*. The Ramblers Association, London.

Ratcliffe, D.A. (1981) The purpose of nature conservation. *Ecos* 2, 8–13.
Ratcliffe, D.A. (1990) Local requirements against national responsibility. *Ecos* 11, 31–35.
Repetto, R. (1988) *The Forest for the Trees? Government Policies and the Misuse of Forest Resources*. World Resources Institute, Washington.
Rockwell, H.W. (1990) Policy update - more on the owl. *Journal of Forestry* 88(8), 11.
Roney, A. (1990) *The European Community Fact Book*. Kogan Page and the London Chamber of Commerce, London.
Shoard, M. (1980) *The Theft of the Countryside*. Temple Smith, London.
Shoard, M (1987) The Land is Our Land: the Struggle for Britain's Countryside. Paladin, London.
Sirén, G., Perttu, K., Sennerby-Forsse, L., Christersson, L., Ledin, S. and Granhall, U. (1984) *Energy Forestry: Information on Research Experiments at the Swedish University of Agricultural Sciences*, SEFP.
Smyth, A.V. (1990) Renewing the conservation ethic. *Journal of Forestry* 88(11), 29–31.
Stewart, P.J. (1988) *Growing Against the Grain*. Council for the Protection of Rural England, London.
Strathclyde Regional Council (Undated) *Forestry Strategy for Strathclyde*. Glasgow.
Thompson, D.A. and Matthews, R.W. (1989) The storage of carbon in trees and timber. *Research Information Note* 160. Forestry Commission, Edinburgh.
Timber Growers England and Wales (1982) *Annual Report 1981/82*. London.
Timber Growers Scotland (1982) *Annual report 1981/82*. Edinburgh.
Watkins, C. (1984) The use of grant aid to encourage woodland planting in Great Britain. *Quarterly Journal of Forestry* 78(4), 213–224.

6

IRELAND

BACKGROUND

Six facts are fundamental to an understanding of the position of forestry in the Irish economy and society.

1. Dependence on agriculture has been and remains high but marginal land, defined as that with major constraints on production, accounts for 49.5% of the agricultural land area (Gardiner and Radford, 1980), and 68% of agricultural land falls into the European Commission's classification of Less Favoured Areas (Bulfin, 1987).
2. Despite a major increase in agricultural output after Ireland's accession to the Treaty of Rome in 1973, agricultural net income has been in decline for some years.
3. Private afforestation increased from an average of 300 ha per year during the period from the 1930s to the early 1980s to 10,000 ha in 1990–1991, the bulk of this being undertaken by farmers.
4. The area of productive forest in 1990 was 430,000 ha relative to a total land area of 6.9m ha, whereas the current scale of new planting (made up of 9300 ha private and 6670 ha state, a total of 16,000 ha in 1989–1990) implies the remarkably high rate of increase of 3.7% per annum on the total woodland area.
5. Given the high rates of tree growth possible and the desire to reduce imports of wood products, the only barrier to extension of forest area has been its financing.
6. Income per head at IR£4950 in 1989 (World Bank, 1990) is lower than in any other north western European country.

Although it is a fact that forest as a proportion of land area is lower in Ireland (6.2% in 1990) than in any EC country apart from The Netherlands, this commonly expressed point has hardly been the decisive one in leading successive Irish governments to support a policy of afforestation and subsequent wood-industry development. The crucial factors are that there is a large potential for wood production which promises to be cheap by Western European standards and that the need to find alternative income and employment for farmers and their families has been and continues to be pressing. Using the agency of the state forest service throughout the post-War period, and with an increasing emphasis towards support of private afforestation since the mid-1980s, Irish governments have continuously favoured the expansion of commercial forest.

In recent years three developments in regional and agricultural policy have been important to forestry and the active agents involved in afforestation. In 1976, the private woodland area totalled only 81,000 ha, of which 33,000 ha was scrub (Bulfin, 1987, quoting Convery, 1980). The main agent of change in an active programme of afforestation was the state. The land concerned was generally poor. The first point to be noted is the attitude of farmers. In 1986, with the belated uptake of the EC-supported Forestry Development Scheme for Disadvantaged Areas (the 'Western Package') to assist public and private sectors in planting an increased programme, the concerns of farmers in one area, Leitrim, were dramatically demonstrated by the burning of £100,000 worth of forestry equipment (Bulfin, 1987). Apprehensions over the loss of government confidence in and declining support for agriculture in the face of the then recent EC-wide supply restrictions and price reductions contributed to this violent reaction. The second development and a most significant change has been the acceptance of the idea of afforestation taking place on better land. An indication of the major change in attitudes to forest development relative to agriculture is the reference in the Programme for Economic and Social Progress (Haughey, 1991) to the afforestation of enclosed and improved land, a move long favoured by foresters aware of the potential of such sites (OCarroll, 1984). The third noteworthy point of more recent years has been the growth in farmers' participation in afforestation. These changes have occurred against the background of little improvement in the economic structure of farming. Indeed the concern to maintain their control over farmland is shown by the tendency for farmers to sell to the state forest enterprise rather than admit defeat by selling to another farmer. Another expression of attitude is the failure of farmer retirement schemes. As a result of both these factors, average farm size has altered only very slightly.

Legislation and Policy

The most important law bearing on forestry dates from 1946. In addition to setting out the functions of the then Forest Service, this Act confirmed the control of felling and restocking on private woodland first introduced by the 1928 Act. However, the Forestry Act is in effect an enabling measure which provides the framework within which new directions of forest policy are operated. The convention is for the responsible Minister to develop policy initiatives and any necessary budgetary steps. In the past decade the latter has implied negotiation of very substantial EC assistance.

The most recent initiative to be implemented is the Forestry Operational Programme 1989–1993 (Molloy, 1990), which sets out the details of the assistance offered from government supported by Structural Funds derived from the European Agricultural Guidance and Guarantee Fund (EAGGF) to both private owners and the state forestry company, Coillte Teoranta. This follows the Western Package 1981–1990 which saw the beginning of a dramatic change in attitudes of Irish farmers towards afforestation.

A statement by the Government (Haughey, 1991), representing the combined views of the bodies linked together as 'the social partners', confirms the intentions of the Operational Programme and sets out a highly specific set of aims as the method of achieving in the forestry sphere the goals of the country's 'Programme for economic and social progress' in the period to 2000. The first four elements of the programme are quoted verbatim below:

1. Expand national planting to 30,000 ha[1] a year to be achieved in 1993 and to be maintained at this level to the year 2000. This will be achieved through a combination of public and private planting.
2. Encourage farmers to further expand their contribution to the planting programme as individuals and through group cooperative structures.
3. The greatest incentives will be available for the planting of lands previously enclosed for agricultural purposes with a weighting in favour of the farmer-owner undertaking such afforestation.
4. Secure EC funds to support the continued development of forestry from 1993 to 2000.

By the standards of wealthier and more urban and populous countries in western Europe, relatively little emphasis was given until relatively recently to environmental, including amenity and recreational, aspects of forest location and operational practices. Major shifts have now been achieved. An indication of the evolution of this aspect of policy was the organization of a conference in 1990 (Mollan and Maloney, 1991) at the instigation of the then chairman of the Irish Timber Growers' Association, B. Hussey. The Forest Service has issued guidelines covering the protection of the environment at large, landscape, fisheries and archaeological sites. As in most countries of western Europe, planting grants are conditional on environmental standards being met. Monitoring of water quality in forest

[1] It should be noted that this figure includes some 4000–5000 ha of restocking, largely by the state forestry company.

areas has recently been initiated. The Forestry Operational Programme drew special attention to the need to encourage broadleaved planting: the intention is to raise the proportion of broadleaves in planting from 2% to 10%. Perhaps the most novel development in policy was the announcement in early 1992 of a grant scheme for Planned Recreational Forests. This aims to encourage the establishment of mainly broadleaved forests and parkland with walks and nature trails, and provision for fishing, stalking and shooting.

While there is an increasing recognition of economic opportunities for the development of tourism potential, the principal motive of government in stimulating afforestation is to provide jobs in order to reduce long-term unemployment and to raise income. Finance is a critical concern as is clear from the government statement on its strategic economic plan. Ireland is engaged in one of the most active 'setaside' programmes (*sensu lato*) in the Western world today. The emphasis in the following discussion is therefore on the means of implementing this programme in the private sector.

Administration

In line with the increasingly frequent arrangement used to administer forestry in western Europe, the Irish government separated the forestry element of the Forest and Wildlife Service into two parts in 1988. On 1 January 1989, the new forestry enterprise, Coillte Teoranta or Irish Forestry Board, was formed as a limited company, the shares of which are held by the Minister responsible for forestry and the Finance Minister. The State Forest Service was made responsible for policy, legislation, private forestry, plant health, research and external relations.

Tax

Taxation of private forestry is highly simplified, the only burden being that arising from capital acquisitions tax. There is no land tax.

Income tax

Woodlands managed by an individual or a company on a commercial basis are not subject to income or corporation tax.

Gift and inheritance tax

The Capital Acquisitions Tax is borne on gifts or inheritances received during the lifetime of the beneficiary, as in most western European countries, and, as elsewhere, the tax rates are based on the degree of consanguinity. The basis of valuation of property passing by gift or bequest is market value, but reliefs are offered in respect of woodland. Where the beneficiary is a farmer as defined above, the taxable value is the open market value less 50% or IR£200,000, whichever is the lesser. For other beneficiaries, the trees but not the land are relieved to the extent of 50%.

Because the burdens differ so markedly with whether the donee is a farmer or not two cases are considered. Macra na Feirme (1992) has shown that 76% of farm operators farming their parents' or relatives' land expect to inherit the farm and death was expected to be by far the commonest occasion of transfer. For an average case the tax rate applying in these circumstances is assumed to be 8%. The value of land for tax purposes is taken to average no higher than 0.5 × IR£400 ha^{-1}, and of growing stock 0.5 × IR£1200 ha^{-1}. Assuming the total taxable value passing to be less than IR£200,000 the tax charge then equals 0.08 × £800 or IR£64 ha^{-1}. If transfers occur every 25 years and the real rate of interest is taken to be 2%, the annual equivalent of this charge is, as explained in Chapter 4, 0.03 × £64, or £2 ha^{-1}. For the non-farmer, and again with transfer to a child but at a higher average rate of 10% and a value of IR£1000, the annual equivalent charge is 0.03 × 0.1 × 1000 or IR£3 ha^{-1}.

Wealth tax

Although a wealth tax applied between 1975 and 1978, no such tax exists at present.

Summation of direct tax burden on private forest owners

Tax payments are estimated as follows for typical conifer crop values: inheritance tax, and total IR£ ha^{-1} $year^{-1}$ 2 and 3 for farmers and others, respectively.

GRANTS

Reflecting their concern with the establishment of well-stocked timber-producing plantations, the current schemes concentrate on capital aspects, namely the formation of plantations, including 'improvement' and 'reconstitution', the building of roads and the purchase of logging machinery. There is, however, an extremely important addition to the schemes under the Operational Programme rubric. The Forest Premium Scheme, also launched in 1990, offers annual payments for 15 years from the year of planting to all those recognized in the scheme as farmers. In 1986, a scheme had been brought into operation whereby headage payments to farmers in respect of livestock could continue to be paid after afforestation. The case for a more universal form of annual incentive payment system was strongly argued by, among others, Bulfin (1987). Apart from the Premium Scheme, little attention is given in the various schemes either to maintenance expenditure, though attention is paid to first thinnings, or to management matters such as the preparation of working plans.

Under the new scheme covering the whole of Ireland and made possible through Structural Funds support, as well as in relation to the Forest Premium Scheme, an important distinction relates to the identification of a 'farmer'. A farmer is defined as one whose annual off-farm income does not exceed IR£13,900 in any of the three years previous to planting, who lives within daily commuting distance of the plantation, regarded as up to 70 miles (*sic*), and who owns the land in question. Part-time farmers and farmer cooperatives also qualify for the scheme. All the rates shown below are those applying from January 1992. Forest Service booklets (1987, 1990a, 1990b, 1992) indicate the main features of the schemes.

Planting

Afforestation

Farmers: up to 85% of approved plantation establishment costs; others: up to 70%. Maximum grants are:

- IR£900 ha^{-1} for conifers on land previously unenclosed for agricultural purposes;
- IR£1100 ha^{-1} for conifers on previously enclosed and improved land;
- IR£2000 ha^{-1} for broadleaves (a 40% rise on the previous level).

Payments are made in two instalments, 75% at planting and 25% after 4 years subject to satisfactory establishment. Minimum block sizes are 2 ha for conifers except where close to or actually adjoining existing plantations, and 0.5 ha in the case of broadleaves. There are no variations in allowable maxima with block size above the minima.

The Forest Premium Scheme, farmers only, including part-time farmers, provides annual payments of:

- IR£50 ha^{-1} for 15 years after planting of previously unenclosed land;

- IR£116 ha^{-1} for the first 20 acres (8 ha) and IR£100 ha^{-1} thereafter for 15 years for previously enclosed land;
- IR£116 ha^{-1} for 20 years on all broadleaved plantations;
- a flat rate of IR£50 ha^{-1} for 15 years for farmers, including part-time farmers, whose income exceeds the threshold;
- an upper limit of IR£6000 per annum per beneficiary.

Restocking

Rates of IR£500 ha^{-1} apply for conifers, and IR£800 ha^{-1} for broadleaves. Payment by instalments and minimum block sizes is as noted above. EC support is not provided for this grant.

Improvement and reconstitution of woodland

Improvement is taken to mean the conversion of neglected woodland to productive forest, whereas reconstitution is concerned with the restoration of woodland damaged by fire, wind or other natural causes. Farmers can receive up to 85% of approved costs, others up to 70%. Maximum grants are:

- IR£1100 ha^{-1} for conifers,
- IR£2000 ha^{-1} for broadleaves.

Forest roads

For all applicants, 80% of approved cost is available, up to a maximum rate of grant of IR£12 m^{-1}. Preplanting roads are allowable.

Harvesting machinery

Up to 45% of investments in machinery for felling, processing and extraction of timber, including development or adaptation costs where applicable, is allowable – a clear subsidy to current output.

'Back-up measures'

Grant aid is available up to 60% of investment expenditure for projects meriting assistance which relate to forest development studies or demonstration projects, or to the formation of associations, including those promoting consolidation of farm forestry (*sic*) in the region.

Planned recreational forests

The conditions for the establishment of recreational forests include a minimum of 10% broadleaved component but are otherwise more flexible than those under the Forestry Operational Programme and allow amenity and recreational features to be included without diminishing eligibility for grant. A further incentive of IR£200 ha^{-1} is aimed at local authorities and other public agencies in order to encourage community involvement.

EXTENSION

Bulfin (1991) emphasizes the importance of an active advisory service in promoting forestry among farmers who have no tradition of forestry. Commercial organizations also play an important role in promotion as their first point of contact with farmers. The Forest Service, together with staff of the national agricultural advisory service, have given much attention to promotion and advice. With the formation of the new Forest Service it was decided that Coillte staff should be employed on an agency basis to undertake the extension and administrative functions concerned with the new and existing grant schemes. However, this arrangement did not prove entirely satisfactory and accordingly from 1992 some 16 staff will be employed directly by the Forest Service (Coillte, 1991). The estimated cost is IR£500,000 p.a. at 1991 rates. A considerable part of the current afforestation is carried out by contractors, thus limiting the call on public extension and advisory services by the individual owner.

RESEARCH

Total expenditure on forestry research in 1990 was estimated (Mulloy, personal communication) at IR£1.85m from domestic resources and a total of IR£2.3m including EC-funded projects. The report of Coillte Teoranta does not identify its research and development spending. It is assumed that the proportion of the total attributable to the private sector cannot exceed 50% and probably realistically can hardly be set lower than 25%. For working purposes the figure of IR£0.8 m is adopted here as the value of resources that can be allocated to private forestry.

TOTAL PUBLIC EXPENDITURE IN SUPPORT OF PRIVATE FORESTRY

Total grant payments are increasing rapidly as a result of the take-up of the various elements of the schemes and will rise both as a result of the accumulating annual payments under the Forest Premium Scheme and the new rates of grant. Because of the changing position, the choice of year in which comparisons of total expenditure on private forestry are made has a marked influence on the result. Approximate estimates of the broad levels of expenditure in 1991 are as follows (IR £m year^{-1}):

- Grants 10.0
- Extension 0.5
- Research 0.8
- Total 11.3

Effects and Implications of Implementation Measures

It is clear that the history of post-war attempts to encourage afforestation by private individuals must turn on the ability of the authorities to persuade farmers that the particular scheme offers something favourable for them. In Ireland, any scheme which fails to recognize the paramountcy of the farmer's, or his heirs', interest in land is doomed to failure. The country's agrarian history, including the Irish farmer's memory of the Famine and the landlords of a previous era, conspires to make conservation of ownership an immovable feature.

The slow initial uptake of major EC-supported measures for afforestation, the Western Package, was due to several factors. In the first place no income support was provided over the early years of the life of a plantation. Second, a forestry tradition was wholly lacking among the landowners at whom the scheme was aimed. Third, there was insufficient promotion among these farmers who were often biassed against forestry. The Forestry Operational Programme, with 70% funding from the European Community, was not only backed more convincingly by government, it was also launched with much more publicity. But, even more decisively, the farmer's position was recognized by offering an annual payment to cover the non-revenue producing years through the medium of the Forest Premium. In 1991 an impressive 71% of private afforestation was carried out by farmers. It has become fashionable to speak of government measures such as grants schemes being virtuous if they are targeted; the Forest Premium Scheme clearly meets that test.

Review of Policy Issues

Two central issues are outstanding features of Irish forestry in recent years. First has been recognition of the fact that forestry offered an important economic opportunity and at the same time faced a particular challenge in assisting with the adjustment of land use from a sector of agriculture, livestock rearing, which was in serious decline, to a potentially more productive use as forest. The second, and of no less importance, was the realization that the business of creating new forest resources need not only be a matter for direct involvement by the state. This latter revolution in thinking, which at first sight appears curious to an outside observer, is intimately bound up with the prevalent attitudes of Irish landowners who are predominantly small owner-occupiers. The importance of this factor can hardly be overstressed.

The massive changes in land use with the expansion of planting have wide landscape and other effects. The Forest Service has a major mission in ensuring that forestry practice should meet high environmental standards. One expression of this is concern over monospecific planting schemes which may appear rational to the individual owner but diversification of which may be desirable for amenity and other reasons. The newly established recreational forest scheme may be expected to help to widen the appeal of forests and forestry.

References

Bulfin, M. (1987) Determining the role of private forestry on highly productive sites in agriculturally disadvantaged areas. Project under the EEC R and D programme: Wood as a renewable raw material; Contract BOS 101 EIR. Typescript.

Bulfin, M. (1991) Private forestry in Ireland: progress and E problems. Paper to EC (DG VI) workshop: afforestation of agricultural land. Typescript.

Coillte Teoranta (1991) *Annual Report and Accounts 1990*. Dublin.

Forest Service (1987) *Grants for Private Forestry*. Forest Service, Department of Energy, Dublin.

Forest Service (1990a) *Forest Premium for Farmers*. Forest Service, Department of Energy, Dublin.

Forest Service (1990b) *Grant Schemes for Forestry*. Forest Service, Department of Energy, Dublin.

Forest Service (1992) *Rural Development Through Forestry*. Forest Service, Department of Energy, Dublin.

Gardiner, M.J. and Radford, T. (1980) *Soil Associations of Ireland and Their Land Use Potential*. An Foras Taluntais, Dublin.

Haughey, C.J. (1991) *Programme for Economic and Social Progress*. Stationery Office, Dublin.

Macra na Feirme (1992) *Land Transfer Survey*. Dublin.

Mollan, C. and Maloney, M. (eds) (1991) The right trees in the right places. In: *Proceedings of a Conference on Forestry and the Environment, 1990*. Royal Dublin Society.

Molloy, R. (1990) *Ireland: Forestry Operational Programme*. Stationery Office, Dublin.
OCarroll, N. (1984) *The Forests of Ireland*. Turoe Press, Dublin in association with Marion Boyars, London.
World Bank (1990) *The World Bank Atlas*. Washington, DC.

7

FRANCE

BACKGROUND

Seven statistics may be noted as being relevant to any discussion of forest policy in France:

1. The forest area is large at 14m ha, or 26% of the land area.
2. The forests are very diverse in terms of type and ownership.
3. Private forests account for a dominant 10.2m ha or 71% of the total forest area and their owners demonstrate a wide range of aims and abilities.
4. 25% of the area of private forests occurs in ownerships of less than 4 ha.
5. Coppice and coppice with standards constitute 29% of the forest area, much of it private.
6. Excluding fuelwood, the cut of which amounts to 10–15 m^3 annually, net imports of wood fibre in 1988 amounted to 10m m^3 (Bazire and Gadant, 1991) relative to removals of industrial wood[1] of 35m m^3.
7. With the growth of tourism, forest fires on the Mediterranean littoral have raised the profile of forestry to an embarrassing degree.

France's richness in forests carries several implications. The first is that there is a strong forestry heritage and forests are in consequence regarded as an important and accepted part of the rural scene. Second, because of their extent, the pressure to maximize the value of output from them is far less than in countries that are either poorer economically or less well wooded.

The third point above emphasizes the importance of private woodlands, almost all of them owned by individuals or occasionally family *groupements forestiers* (see below).

Fourth, the number of small woods as viewed from a more traditional

[1] The term industrial wood is used in the sense employed by FAO.

forestry perspective suggests a management problem. As such it has been a focus for a vast array of legal and financial, including fiscal, initiatives, as well as major extension efforts over the past 45 years. These have been erected on the basis that a small scale of management unit is necessarily bad. This may be true in commercial terms; harvesting costs rise with small quantities and even high quality material tends to be bulked with low. Nevertheless, such small woodlots represent, despite the denigrations of timber producers, a valued resource still largely exploited for wood used locally in assortments ranging from fuel to poles and fencing to sawtimber. But such woodland also contributes massively to the landscape diversity and wildlife richness of France's countryside. Given its geographical distribution, often in small parcels, it is indeed questionable whether the bulk of this woodland could be used and enjoyed more economically.

Coppice and coppice with standards have long constituted an anachronism in relation to 20th century wood needs and harvesting technologies. Coppice types often occupy quite fertile sites and these crops have been subject to many attempts to convert them into more valuable stands: their continued presence constitutes a standing reminder of the conflict between aims and economic realities in forestry.

Sixth, despite its large forest estate, France is a net importer of industrial wood and its products. Because of the landscape security provided by forest and the nature conservation values that it enshrines, it is possible as a broad generalization to say that there is little pressure to alter the condition of the forest on these scores. The aspect of forestry that has caught the attention of policy makers is the wood 'deficit' which is regarded as bad on the familiar autarchic ground beloved of foresters and, it is argued, is remediable at an acceptable cost. It has of course to be remembered that improving the marketing of wood and providing incentives to achieve both this aim and the establishment of more forest act to increase the value of private growers' assets.

The final point about fire and the public passion it arouses is not developed further except to note two aspects. One is that fire damage is an important feature of French forestry not only in the Mediterranean but also in other areas, notably Aquitaine. The second is that the importance of a subject which would otherwise not call for more comment than the ravages caused by so many biotic and abiotic agents lies in the fact that much of the concern does not arise from the loss of forest. The Mediterranean problem stems from the fact that the forest provides the fuel but not the wherewithal in terms of marketable products to justify spending by owners on the forest's protection (Mazodier, 1989; see also Bazire and Gadant, 1991). The case is a classic one of an external diseconomy.

Ownership and Physical Features

One of the ideals of the Revolution, equality, led in 1804 to the enshrinement in the *Code civil* of the idea that all heirs had equal rights to their parent's property. The inheritance law, styled by some 'a machine to cut up land', combined with most landowners' desire to retain land as an asset explains the preponderance of small ownerships so apparent in private woodlands ownership. Surprisingly, the number of farms, which also increased in the 19th century, reached a peak of about 3.5m in the 1880s (Lowe and Bodiguel, 1990). The same does not appear true of private woodlands. Buttoud (1984) suggests that almost all the increase in area of some 3m ha between 1910 and 1980 was in woodland ownerships of less than 50 ha. Part of this unbalanced shift was no doubt due to the splitting of properties through inheritance and part to the postage-stamp planting encouraged by the Fonds Forestier National (FFN), which often included planting of abandoned fields.

Clerc (1980), basing his figures largely on those of Buttoud (1979), identified four distinct categories of private forest owner:

1. The most important class, holding approximately 100 ha on average is long established and occupies 3.1m ha or one-third of the private forest area. (Note that Ministére de l'Agriculture, 1987 shows 2.41m ha relative to 9.738m ha, the surveyed area of non-public forest, or 25%.)
2. Farmers who own 4m ha of land with at least one wood.
3. Town dwellers who, by inheritance or purchase, own 2m ha.
4. Financial institutions (insurance companies, banks) which account for 150,000 ha[1].

Reed (1954), commenting shortly after the end of the Second World War and before the emergence of the problems facing forest planners 45 years on from the start of the work of the Fonds Forestier National, could write of post-Revolutionary France, and with percipience in view of the situation that has actually developed,

> Rights of property being inviolate, privately owned forests remained immune from State control, and in this age of laissez-faire the results were not altogether happy. . . . Private woods, unlike those in communal possession, are not subject to the provisions of the Forest Code and private property rights having remained (thanks to the French individualist tradition) respectable[2] and, indeed, sacrosanct the coercive powers enjoyed by the State are strictly limited.

---

[1] Guillard (personal communication) estimated that in 1991 these amounted to 200,000 ha.

[2] Meaning, presumably, 'respected'.

In the 1950s the State, as Reed pointed out, had control over private woodlands only in three contingencies and in two of these the surrender of private rights was voluntary. The two he noted concern first, forests made subject to Forest Service administration (loi Audiffred) when a liability for payment for services rendered accrues although such woods were then partly exempted from succession taxes, and second, forests restocked by the Forest Service with the aid of a loan from the Fonds Forestier National. The other case concerned protection forest to which some of the earliest regulations and public actions attended. As explained below, several new laws, institutions and burdens have been introduced since Reed was writing, each of which has led to some, not very marked, qualification of the freedom of owners to act in their own interest. Principal among these has been the progressive extension of controls over deforestation. These originally embraced forests with a protective function and which now extend effectively to all woods over 4 ha in extent for the sake of 'the biological equilibrium of the region' and most recently to the purpose of providing for 'the well-being of wildlife'.

Overall, the average growing stock per hectare of private forests is substantially lower for both broadleaves and conifers than in State forests, whereas the figures for current increment are very nearly the same (Bazire and Gadant, 1991, Table 10). However, the sheer scale of forest cover over France's widely varying geography means that aggregate statistics obscure much regional variation. It is not safe to assume that site for site and age for age private owners hold lower growing stocks than the state forests, though this would not be surprising in view of the past conservative nature of the management of the latter. The proportion of private forest within the whole ranges from 96% in Limousin to 32% in Lorraine-Alsace (ANCRPF, 1991) and no doubt similar variation exists in other parameters measuring the characteristics of the growing stock and its management. Coppice in non-Mediterranean areas often occurs on sites capable of producing a far higher value of wood per hectare under high forest. This fact is enough to make this form of cover the target for much effort in so-called improvement.

Although the private holding of forest contains high proportions of both the total area of mountain forest and the Mediterranean cover types which are increasingly regarded as important, the dominant interest in forest management as expressed by the private owners' organization and government is the production of wood in the form of sawlogs. Guillard (personal communication) gives as reasons for this emphasis in the past, especially at the start-up of the Fonds Forestier National, that sawlogs are more valuable than small roundwood, and that pulpwood supplies were satisfactory; a clear demonstration of the economic rationale is provided by poplar cultivation which is dominantly in private hands and which accounts for 10% of broadleaved timber production from only 1% of total forest area. This 'productivist' aim, paralleling that for agriculture (Lowe and Bodiguel, 1990), follows a traditional line of autarchic argument owing little to economics. This runs from recognition of a substantial dependence on imports of processed wood goods to the plausible sounding but unconvincing line of argument that puts the reduction of import dependence as the dominant aim, with attention to soil protection functions of the forest in certain regions, and a ceremonial bowing to the needs of recreation and occasionally landscape. This general argument, which underlies forest policy as a whole, is particularly favoured by the private sector of forestry and by the (private) sawmilling industry.

The sawmilling industry is viewed as a key, but currently inefficient, partner in the forestry industry as a whole, yet relatively few studies have been made of its functioning, economic efficiency, etc. (though Levêque and Péguret, 1988, have helped to fill the gap). A clear demonstration of the improvement achieved in technical aspects and implicitly in the cost efficiency of modernization schemes is given by Parrat and Hocquet (1976).

The trade deficit in pulp and paper greatly outweighs that in mechanical wood products, even if this is taken to include furniture. The difficult situation concerning the total scale and poor coverage of pulping capacity of the 1970s and early 1980s has been relieved as a result of major investments, mainly by foreign pulp companies. However, but the concern about the ability to match potential wood supply and use of small roundwood by industry remains, as shown by the report of an industry strategy group set up by the Planning Commission (Guillery, 1991).

As in all forests there is a high interest in hunting in French private forests. No comprehensive data on the returns from hunting are available but the economic values associated with the activity are no doubt substantial; for example in some areas sporting rents cover the land tax. Equally there is little doubt about the importance attached by society to the forest's contribution to general access for recreation. Public funds can be granted to owners opening their woods. Specific provision for wildlife conservation is not a main concern of French forestry. The references to protection of the country's heritage are in most cases to the forest as a whole rather than assemblages of particular habitats or groups of organisms.

Legislation

The *Code forestier*, a compendium of laws and decrees bearing on forestry first issued in 1827, combines in one volume all relevant current legislation. A valuable panoramic view of the important laws, etc. is provided by Lagarde (1991).

Certain very specific legal requirements are placed on private woodland owners. Where a law expresses intentions rather than obligations, however, implementation may be half-hearted; this applies for example to the section of the 1985 law concerning consolidation of holdings. The following laws are important.

1. In respect of protection forest: the laws of 1827, the year of the first *Code forestier*, of 1882 concerned with protection forest in the mountains, and particularly of 1922 although this last was not actively applied and the law of 1976 dealing with suburban and tourist forest areas which was used to reinvigorate the requirements of the original law.
2. Control of deforestation: this is relaxed for areas under 4 ha, woods under 20 years of age, and parks and gardens of less than 10 ha. The justifications for control elsewhere than in these cases lie in concerns for soil protection, dune and watershed protection, protection of investments aided by the State, and, more recently as already noted, wildlife conservation. In addition to the prohibition that applies unless clearance is approved by the Prefect, a further disincentive was introduced by laws of 1969 and of 1985 when specific penalties were established, with revisions and clarifications enacted in 1990 as to the persons on whom they should fall.
3. Management plans (*plans simples de gestion* or PSG): these were introduced along with the other major changes in the administration of forestry in 1963 as obligatory for all private forestry properties above a certain size; the general rule applying to a property containing at least 25 ha in a single block, or for coppice 40 ha. Young plantations pre-thinning stage are exempt. The form of plan is quite simple and, if anything, less formal and less used managerially than the plans of operations of the old British Dedication scheme. The existence of a plan relieves the owner of the need, *inter alia*, to obtain specific permission for a clear felling. Owners with between 10 and 25 ha may produce a plan: the incentive for this is a reservation of public funds for such owners' needs (Bazire and Gadant, 1991, Chapter 9). In 1989 it was estimated that relative to an area of 3.48m ha for which PSG were obligatory, active plans existed for 2.4m ha and a further area of almost 1m ha was worked under a system of administrative authorization of felling.
4. Finally, obligations set out in the law of 1985 for the management of vegetation in regions of high fire risk.

Policy

Formulation

Britain conducts its debates on forestry policy in what may be considered to be an authoritarian and indeed sometimes adversarial manner. The

Government proposes new principles and/or plans and the reaction emerges later, and perhaps too late for the current round of policy revision. A technique commonly adopted in France is to appoint some senior public figure such as a back-bench MP or a few experienced professionals to prepare a report. This carries no official imprimatur and thus provides the government with room for manoeuvre in defining a policy line in due course. Examples of the recent past are de Jouvenel (1978), Betolaud and Meo (1978) and Duroure (1982).

In de Jouvenel's (1978) report arguments were advanced for the provision of greater powers to foresters and for a grand policy for forestry. Guillard and Pardé (1976) edited a report on wood resources and requirements which gives useful background to the report on the 'wood chain' by Betolaud and Meo (1978). In 1981 the Prime Minister, P. Mauroy, charged an MP, R. Duroure, with the task of preparing a report to cover specifically the elements of a new forest policy, the conditions necessary to establish a better marketing system for wood, a topic perceived as being of particular importance, and the organizational arrangements needed to effect the desirable changes. The report assesses the costs but does not quantify the benefits of the programmes and level of State aid recommended. A principal concern lay with the need to harvest more of the wood coming on stream from the extensive and often scattered post-war plantations, to utilize it in higher value forms and so to achieve the much desired reduction in the trade deficit in wood and wood products. A strong recommendation was entered for the formation of an autonomous authority for forestry and forest industry, a view which found no immediate echo in government although, for one year in 1985–1986, a secretaryship was created under the Minister of Agriculture to deal solely with forestry matters. Though it had only a short life, this has been quite effective in stimulating investment in wood industry.

The reader is referred to the particular reports for a full assessment of their coverage and depth. Their most noticeable features are the dominance of technical argument, the low economic content, features surprising for a country which prides itself on its economic achievements, and the lack of attention to conservation and recreation goals (although de Jouvenel's report devoted space to the 'social role' of forests), itself no doubt a consequence of the same prevailing attitudes towards the generation of profit in a liberal economy. These aspects may be taken as an indication of the strength of the forestry lobby, which is hardly well-known for its encouragement of critical economic appraisal, and creates the suspicion in the reader's mind that the principal actors concerned have always been fully aware of the financial dimension but unwilling to clarify its significance.

Public attitudes towards forestry

As already noted, fires in the maquis of the Mediterranean littoral have caused great public concern in recent years. This arises not so much because of the damage caused to the vegetation as for the risks to neighbouring property. It is ironic that the mass of combustible material has risen with the colonization of abandoned agricultural holdings by shrub and tree species; forests are thus at the heart of the cause of the risk.

One ever-present concern, even of the general public, is the choice of species either in restocking or, more obviously, within afforestation. A number of steps have been taken to encourage broadleaves in planting. On the other hand very little attention is paid to landscape design. More recently, ecologists, or, more accurately, conservationists, have raised objections to the use of herbicides, and the impact of forestry on particularly valued habitats. It is sadly true, as Martres (1991) points out, that too often the critics speak in ignorance of the biological facts.

Apart from these concerns, which show no sign of significantly altering in their relative importance, the dominant forces influencing private forestry policy are internal to the forestry sector. They are first, the private owners themselves, and within that group those with larger properties, second, the wood-processing industry with emphasis on the sawmillers, and third, forestry technocrats concerned especially with the inefficiencies implied by small-scale ownership and the removal of impediments to economically efficient management.

Current policy

There is no single statement of forestry policy. The *Code forestier* compiles the current laws and decrees bearing on forestry, including those on taxation. It should be remembered that the existence of a law does not imply that it will always be implemented; this may be especially true of France.

One of the most recent laws (4 December 1985) introduced a number of initiatives or refinements of earlier legislation. One departure stated that the policy for enhancing the economic, ecological and social value of the forest was to be transferred from the competence of the state and passed to the Commissions Régionales de la Forêt et des Produits Forestiers (CRFPF). Their plans, entitled 'Orientations régionales forestières (ORF)' include broad statements of policy for forestry and forest industry and are developed in 'directives' for State forests of the region, 'orientations' for the communal forests and 'orientations régionales de la production'. By 1991 ORF had been prepared for most of the 22 regions. They have to be approved by the Minister of Agriculture. They may contain financing plans for regionally important investments. There is no definite frequency set for the plans but a ten-yearly revision is foreseen.

Afforestation may not be undertaken in certain specified localities, principally for reasons of enhanced fire risk. A decree dated 6 December 1991 established a grant for the afforestation of agricultural land. Much afforestation of abandoned farmland has of course been undertaken over the past 44 years through the agency of the Fonds Forestier National. It should be noted that the *Code rurale* provides that communes can forbid afforestation of defined lands (Ministère de l'Agriculture, undated).

Both fiscal and grant incentives are designed to favour restocking and afforestation and strict controls ensure that clear felling is followed by restocking. Although the French contribution to FAO's survey of forest policies in Europe (FAO, 1988) refers to the value of the forests of the state and local authorities for wildlife conservation and recreation it is silent on the roles of private woodlands in these areas.

Administration

The law of 1963 made three important administrative changes to the former organization of the Département des Eaux et Forêts. In the first place a forest authority was set up. As of 1991, the Sous-Direction de la Forêt constituted the forest authority. Its responsibilities are those usual to such institutions namely; development of policy and control of its application; forest management and protection; the marketing and promotion of wood; forest land use planning; participation in the organization of research, training and education; management of the Fonds Forestier National; supervision of the state forestry enterprise and of the Centres Régionaux de la Propriété Forestièrè described below. The organizational arrangement is that the Forest Authority forms part of the Direction de l'Espace Rural et de la Forêt (DERF) of the Ministère de l'Agriculture et de la Forêt. Secondly the forestry enterprise was established as the Office National des Forêts. This was put under the aegis of the forest authority, and was subsequently given greater autonomy in return for it arranging its affairs so as to be self-financing.

The third change was the creation of an extension service in the regions provided by 17 Centres Régionaux de la Propriété Forestière: these came into being in 1966. Their functions and activities are discussed more fully below. So far as private forest owners are concerned, the CRPF checks management plans and gives advice, while the authority carries out its administrative functions such as the control of felling and payment of grants through the Service Régional de la Forêt et du Bois (SERFOB), a special branch of the DDAF or Direction Departementale de l'Agriculture et de la

Forêt. The DDAF is in turn managed by the Direction Régionale de l'Agriculture et de la Forêt (DRAF).

A total of 76 Syndicats de Propriétaires Forestiers Sylviculteurs in the Départements are grouped into a Federation Nationale for the representation of private forest owners' interests. The central organization is an influential body, and maintains a close eye on the working of the CsRPF and the National Association of CRPF. It produces the journal *Forêts de France et Action Forestière*.

Implementation Measures

The earlier discussion of important laws listed a number of initiatives and controls. So far as the former are concerned, these may remain ideas or good intentions never to be implemented. Others, such as those setting out obligations on restocking after clear felling, demand action. But all that is written in decrees and laws is not to be taken as necessarily calling for action. In addition to the legal framework, three main elements guide the practical application of forestry policy in France, namely:

1. Financial assistance from the Fonds Forestier National.
2. Various organizational measures, principally groupements, assisted by the Fonds and designed to gain scale economies in silvicultural or harvesting and marketing operations.
3. The extension system embodied in the Centres Régionaux de la Propriété Forestière.

Groupements

Because groupements play such an important role in French forestry it is useful to outline their purposes and forms. (The French word is retained throughout this section because the terms used are highly specific to different forms of entity, each with their own legal and financial implications, and because there are not always exact English equivalents).

There are two categories of groupement, one of which involves a banding together of ownerships. This is the groupement forestier which is a private company. A total of 720,000 ha were held in this form of tenure in 1991 (Bazire and Gadant, 1991) under which the owner's independent ownership is lost. The effectiveness of the institution may be doubted since only 600 out of 3600 of these groupements were formed by small owners possessing small parcels of land. Second, as Buttoud (1984) points out, two-thirds of them involve consolidation under one management of family properties brought together in this fashion largely for the sake of improving the inheritance tax position. In practice, such groupements have little attraction for others than those with close family ties because of the loss of individual ownership.

The other category of groupement is a more or less formal association of owners or operators for purposes of management. Dubois and Normandin (1985) used a questionnaire survey to derive more information on the achievements of such cooperatives which they term groupements de service. The original term used, prior to 1985, was groupement de gestion. These organizations deal largely with silvicultural affairs and work for the common good of a number of owners without involving any transfer of forest property. This is the organization which was favoured for much planting and has achieved good success with grouped wood sales. Substantial numbers of owners (5000) and hectares (600,000) are involved, but Buttoud (1984) suggests that despite its relative success the scheme may have reached the limit of its appeal.

Since 1985, the term groupement de gestion has become a generic term for all such institutions which differ in their function, constitution and method of working. The various forms of these groupements de producteurs forestiers are intended to assist with one or more of all aspects of wood production from silviculture through to marketing of forest products. Staff support can be provided through the Centres Régionaux des Propriété Forestières for this form of society, which however has no special funding and which in fact remains a concept which has proved ineffective. On the other hand 'cooperatives pour l'utilisation du machines' and 'cooperatives de vente' play an important role in mobilizing wood. These are groups largely created by foresters themselves since 1975, and are the commercially relevant agents.

Another new concept, that of the Association Syndical de Gestion Forestier, should also be mentioned. This is a rather special form of cooperative since management decisions taken by a majority of members must be adopted by all. Understandably very few of these organizations have been formed.

It might be thought that the establishment of large-capacity mills must depend on the creation of one or another form of groupement in order to enhance the supply of wood negotiated with larger owners, public or private. In fact the decision to invest in two new pulp and paper mills built in the mid-80s hardly depended on the aggregation of many small supplies into an organized whole. As in Britain, small suppliers climb on to the band-wagon; in effect the market, which includes the harvesting contractors and round timber merchants, is quite efficient in bringing these supplies into play.

Finally in this connection a public institution should be noted. *Regroupement par acquisition* is a method of purchase by public authorities in order to improve the management prospects for small pieces of land: it is little practised despite its immediate attraction and the relatively successful example of SAFER (Sociétes d'Amenagement Foncier et d'Etablissement Rural) in the agricultural domain.

CRPF

There are 17 Centres Régionaux which are the forestry extension arms of government. Their aims as set out in the law are:

1. To assist in meeting the nation's needs in developing the production, harvesting and marketing of forest products.
2. To encourage regrouping, especially by cooperative action, to improve the quality of wood and its products and to enhance the profitability of silviculture.

The principal functions in achieving these aims are listed below. Though the emphasis on wood production is again notable, the functions are a valuable reference point since they provide a useful description of what any competent extension service should do, namely:

1. To disseminate silvicultural, management and marketing expertise.
2. To monitor and assist in planning wood production in the region.
3. To approve management plans.
4. To develop associations of woodland owners and to foster cooperation between landowners in forest management and wood marketing.

Four features of CRPF are noteworthy. The first is that much of the strength of these bodies derives from the fact that they have to check management plans. These are important because of the management and tax benefits that an approved plan confers. The second lies in their being regional. The third is the engagement of forest owners in the region: two-thirds of the governing council of each are drawn from owners and all the council members are elected by forest owners. Fourth, there is an unrepentant emphasis on profitable wood production in the terms of reference and the actual efforts of the CRPF on the ground.

Tax

Cadastral income

Two forms of taxes are levied on income or putative income. Since both are based on 'cadastral income' it is useful to describe the meaning and derivation of this concept. (Mollière and de Reure, 1988 and 1989, provide an invaluable guide to the intricacies of French forestry taxation.) There are 13 different categories of unbuilt land of which one is entitled 'woods and forests'. This category is divided into eight subgroups (broadleaved high forest, coppice with standards, etc.). Each of these is further subdivided into classes which depend on the quality and (physical) productivity of the particular stand. The cadastral income is fixed for all types of landed property at 80% of the '*valeur locative*'. This corresponds to the annual rent that an owner would normally receive if the property were let and is itself set at 90% of the net product of the property, 10% being deducted as an allowance against the risk borne by the owner and his management input. In forestry the net product is calculated as the net annual income expected from the particular stand or a standard version of it. In principle the calculation should be carried out every 6 years, but in practice annual adjustments are made on the basis of information on changes in wood prices. Average volume yields are assumed to remain constant. A source of worry in assessing net incomes is the allowance made for the costs of protection, maintenance and restocking; the 1990 decree establishing new provisions for these costs in preparation for the revision of the cadastre has caused concern (Leclerc de Hautecloque, 1991).

Income tax

The cadastral income is aggregated with an individual's other income. Obviously under the arrangement using a standard net annual income approach no reference is made to the age of the actual stand and the owner's ability to pay out of forest income. The method adopted in France is to relieve young stands entirely, thus: poplars for the first 10 years, conifers for 20 and broadleaves other than poplar 30 years. Special provisions are also made in the case of storm damage to reflect the fact that the basis of a calculated normal average net income has been disturbed.

For purposes of calculating an average assessment to income tax, it is assumed here that cadastral income averages 200 Fr.fr. ha^{-} $year^{-1}$, and that the average rate of income tax for a typical owner of the average-sized forest property is 30%. Allowance must be made for the relief on young stands, that is by some 25%, this being the proportion that the years noted above bear to average rotations in France. The net result is $0.75 \times 0.3 \times 200 = 45$ Fr.fr ha^{-1}.

Land tax

Land tax forms the basis of local revenues and also those designated for certain public institutions. The actual burden is based on the rate fixed by the local authority and applied to the cadastral income. The result is usually low and not dissimilar from the level of the former Schedule B of income tax in Britain if that had been uprated to take account of inflation since its cancellation in the 1970s. An indicative figure is of the order of 60 Fr.fr. ha^{-1} in 1991 for high forest and much lower for coppice (de Reure, personal communication). Another source (FAO, 1988) states that the land tax for State forests, the area of which in 1985 was some 1.432m ha, was 117m Fr.fr., implying an average for these properties of about 80 Fr. fr ha^{-1}. The level may however be substantial if the commune has few other sources of land tax and/or is required to raise increased amounts. Relief is given on young plantations, whether new plantings or restocking but excluding natural regeneration, up to an age of 30 years. This exacerbates the difficulty for communes of raising tax revenues where active afforestation is being undertaken since the tax base can thereby be severely reduced (Buttoud and Rochot, 1977).

Assuming the full rate to be 60 Fr.fr., and the effect of relief to reduce this by one-third, the average land tax borne becomes $0.67 \times 60 =$ 40 Fr.fr. ha^{-1}.

Wealth tax

Wealth tax is a less permanent feature of the fiscal scene, its appearances varying with the complexion of the government. Thus it was introduced in 1982, repealed in 1986 and reintroduced in 1991. The rates of tax are as follows: zero for net worth less than 4m Fr.fr.; 0.5% for 4–6.5m Fr.fr.; rising to a maximum of 1.1% for fortunes over 20m Fr.fr. The basis of the charge is on the soil only, but there is an important exemption, namely where the forest is the source of the person's livelihood no tax is levied. There are rules determining the bona fides of those claiming exemption: how far these succeed in avoiding the manipulations common with capital taxation remains to be seen. For purposes of this chapter it is assumed that the forest is of an average size and that it is owned by a person whose net wealth lies below the threshold.

Inheritance tax

Rates of transfer tax vary with the degree of consanguinity of the beneficiary to the donor. The important provision favouring forestry is that there is a reduction of the forest's value for purposes of calculating the transfer charge, conditional on the new owner's commitment to undertake continued management of the property disposed. This arrangement is called the Serot–Monichon regime after a Minister (Serot) who in 1930 introduced relief for good practice on forest property and a senator (Monichon) who in 1959 introduced the provision that relief should apply to gifts and inheritances. The concept behind the regime is that it is inappropriate to levy a tax on the trees since in effect it implies an additional tax on revenue; but it is considered reasonable to levy a tax on the soil. It is difficult to value the soil separately so the assumption made is that it comprises only one quarter of the total. Thus the capital on which tax is payable is reduced by 75%. In addition, this charge may be deferred by 30 years conditional on the owner pledging himself to good management, this being achieved conveniently by means of a management plan approved by the CRPF. (It has been pointed out by Peltereau-Villeneuve (1991) that the arrangement carries the penalty that it constitutes a barrier to sensible exchanges or groupements.)

The effect of the last provision is that if an heir owns a forest for 30 years before his death, the inheritance tax due is zero. This is the position assumed to apply here.

Summation of burden of direct tax on private forest owners

The average tax payments illustrative of the burden on an average woodland owner carrying on conventional forest practice according to an approved working plan are thus estimated as follows (Fr.fr. ha^{-1} $year^{-1}$);

- Income tax 45
- Land tax 40
- Wealth tax 0
- Inheritance tax (30 year generation) 0
- Total 85

Fonds Forestier National

The Fonds (or FFN) was set up in 1946 in the immediate aftermath of the Second World War with the main aim of reducing the country's dependence on imports of coniferous pulp. An early description of its basis and activities is provided by MacGregor (1954). Its broad purposes include those of improving, protecting, and restocking forests, of afforestation and, significantly, of furthering the marketing of forest produce. A major cause of concern has long been the lack of markets for small roundwood - pulp mills, where present at all, having usually been located near ports and thus too distant for commercial thinning. Aid is available to all categories of forest owner, but under progressively more stringent conditions. Since 1947, planting and restocking assisted by it has amounted to 1.8m ha, 1.4m ha of which are in private forests.

Originally the concept of the FFN was that by being a fund internal to the forestry sector it could isolate itself from variations in government provision of money. An annual account of the financial activities of the fund (DERF, undated) shows the complicated book-keeping associated with this semi-autonomus institution. In the light of what the traffic could reasonably be supposed to bear, the FFN was set up and financed by a tax on the sales of sawlogs, and on home-grown sawnwood. The rate in 1990 was 4.7%. As a result of objections by the European Commission, this cosy system was to be changed in 1990. In the minds of many foresters the method of levying funds adopted appeared fair and appropriate. The charge bore on the products of an industry which it was intended to aid. No burden was placed on other consumers or taxpayers. But this reasoning totally ignores the distorting effects of a method of financing which implied, and indeed still implies, that one class of products made in France is indirectly taxed differentially compared with other products. This constitutes a distortion of trade and is accordingly unacceptable in a Community whose object is to ensure as level a playing field for all economic activity as possible. From 1 January 1991 charges were to be levied on all products of the wood-processing industry consumed in France, at 1.5% on prepared sawnwood, 1% on rough sawnwood, veneers and plywood, 0.5% on joinery products, boxes, particle board, and fibre board, and 0.1% on paper and paperboard (Ministère de l'Agriculture et de la Forêt, 1990). Most important of the revised arrangements was the edict that the resources so mobilized, estimated at 520m Fr.fr. in 1990, should not be spent on other than general aims of forest policy instead of the former arrangement whereby some of the Fonds was devoted to assisting with the improvement of sawmills. Further, the BAPSA tax (du Budget Annexe des Prestations Sociales Agricoles) would no longer be levied in addition to the FFN levy on sawnwood, but on industrial roundwood at roadside in the forest at a rate of 1.3%.

The means of assistance provided from the Fonds are of three forms: materials, cash grants and loans. The arrangements for the major forestry operations set out below are drawn from DERF (1990) and the amounts provided in 1988 from Bazire and Gadant (1991). Since 1985, aid is granted only if the owner undertakes not to fragment his property.

Working plans (plans simples de gestion)

1. Condition: plan to be agreed by the local Centre Régional de la Propriété Forestière.
2. Grant: 20–50% of cost, based on a fixed element of 1000 Fr.fr. and a variable element of 35 Fr.fr. ha^{-1} for the first plan, revisions being paid only on the variable element.

New planting and restocking (apart from natural regeneration)

1. Conditions: operations covered include preparation of the ground, drainage, up to four maintenance operations and access roads and tracks up to 25% of the maximum investment approved, plus consultant's fees. Expenses of protection against mammals allowed where such control still allows the investment to be economically acceptable. Minimum area of 10 ha required for separate woods, otherwise planting must form part of an existing block of 10 ha or be such as to make an existing block attain at least 10 ha. (The minimum area requirement was introduced in order to avoid 'postage stamp' plantations, the marketing of wood from which was seen to be problematic.) Minimum area per species 1 ha except for walnut, cherry, sycamore, red oak and ash for which 0.5 ha is allowed. Minimum expected yield 5 m^3 ha^{-1} $year^{-1}$ or, in the case of poplar, 10 m^3. The sum invested in the case of restocking must exceed 20% of the value of the final cut. Specifications in matters such as required numbers of plants per ha are set out in DERF (1989).
2. Aid:
   (a) plants: free, or one-half cost in the case of poplars, for plants from an approved nursery;
   (b) grant: at rates usually between 20% and a maximum of 50% and averaging 40% of the approved cost, with a maximum of 2000 Fr.fr. ha^{-1} for animal protection;
   (c) loan: at a nominal rate of simple interest of 0.25% for 30 years for an amount of up to 100% of the cost, but averaging 70% of total cost, with the maximum expense allowable for animal protection being 4000 Fr.fr. ha^{-1}.

Conversion of stands by natural regeneration or storing of coppice

1. Conditions: operations allowed in natural regeneration include preparation of ground, singling of seedlings, weeding, removal of coppice and enrichment. Access roads and tracks, firebreaks, drains allowable. The sum invested must exceed 20% of the felling coupe or value resulting from the coppice cut. For natural regeneration, the operation must be conducted on at least 10 ha, on which 70% cover must be obtained within 12–15 years. With storage of coppice, the operation is to be on an area of more than 5 ha with the lot of any sole occupier being at least 1 ha and the yield expected to exceed 5 m^3 ha^{-1} $year^{-1}$.
2. Aid: grant only of between 20 and 50%, with an average of 40% of agreed cost.

Pruning

1. Conditions: applicable to stands which meet the criteria for aid from the Fund. For work on conifers and poplars only, with not fewer than 200 stems ha^{-1} treated, minimum length of clear bole 6 m, maximum 6 m for conifers and 8 m for poplars. Maximum of two prunings aided per stand.
2. Aid: grant only of 20–50%, average 40%.

Installation of a road network

1. Conditions: to cover building and improvement but not maintenance of roads, tracks, depots, turning places. Profitability of project must be sufficiently high. Priority given to large under-roaded areas needing access for mechanized equipment for maintenance of stands as well as opening up for marketing purposes. Investor to provide bond or other guarantee to the value of the loan.
2. Aid:
   (a) grant: 20–50%, average 40%.
   (b) loan: at a nominal rate of 2.5% simple interest up to 100% of agreed cost, but averaging 70%.

Fire protection works

1. Conditions: covering access roads and tracks, fire breaks, water points, fire towers and signs.
2. Aid: grant only of 20–50%, average 40%.

Formation of a group (for obtaining scale economies in forest management and harvesting)

Grant only: 250 Fr.fr. ha^{-1}, (unchanged since 1981), plus 250 Fr.fr. ha^{-1} below 4 ha, 140 Fr.fr. ha^{-1} from 4 to 10 ha, 70 Fr.fr. ha^{-1} above 10 ha up to a maximum of 3450 Fr.fr.

Setaside

The decree of December 1991 set out the terms of assistance offered to farmers, tenants as well as owner-occupiers, who carry out afforestation of arable land. In addition to the FFN grants noted above, and on undertaking an obligation to maintain the plantation for the succeeding period, the farmer receives 1000 Fr.fr. per ha per year for 10 years in the case of fast-growing species (conifers, poplar, walnut) and 15 years (other broadleaves).

Summary

Overall the level of assistance allowable is around 40%, though this does not mean that this average proportion was actually paid. The absolute amount of assistance to private forest owners in 1988 was as shown in Table 7.1 based on Table 14 of Bazire and Gadant (1991).

In addition, expenditure directly related to forestry, as opposed to industry, was made in support of harvesting contractors.

In 1989, the average support for planting and restocking for all categories of owner aided by the Fond was 6420 Fr.fr. ha^{-1} in respect of grant aided work and 10,230 Fr.fr. ha^{-1} in respect of loans (DERF, undated). These sums may be related to costs of establishment per hectare of some 14,000 Fr.fr. ha^{-1}, notably lower than British average costs. For roads the aid granted was 21,760 Fr.fr. km^{-1}.

The bulk of the planting and restocking in France is supported by the Fonds. One commentator (Wheare, 1990) has suggested that the Fonds is not attractive because the terms of the rules employed by the Fonds are onerous, there is an insistence on use of silvicultural techniques, notably relating to initial spacing, determined by the Office National des Forêts, and a low grant payment is given said 'typically to amount to only 25% of the cost of establishment'. Although there are cases of such planting outside the Fond, this is not common practice. The need for standards was recognized in the early 1950s when a multiplicity of dubious techniques were in use. These standards were set by the forerunner of the present Centre national du machinisme agricole, du genie rural, des eaux et des forêts (CEMAGREF) and were made more flexible in the mid-1970s when there

Table 7.1. Analysis of credits provided by the FFN in 1988.

	Totals for individuals and groupements	
Operation	m Fr.fr.	% of total
Planting, restocking, conversion	119	81.5
Roads	22	15.1
Improvement of management	3.4	2.3
Fire protection	1.5	0.1
Total	145.9	100

was a move towards the establishment of more broadleaves (Guillard, personal communication). To British eyes the current technical prescriptions set out for silvicultural operations (DERF, 1990), while arguably on the conservative side, seem eminently reasonable.

Extension

Expenditure is largely that of the CRPF although in addition some work is performed for farmers by foresters employed by Chambers of Agriculture. The mission of the CRFP is clear and the service carries it out with enthusiasm. In 1990, for example, 170,000 people were contacted at least once by a member of CRPF staff and 800 meetings were held, attended by 25,000 people (ANCRPF, 1991). A total of 1650 new management plans were agreed in 1990 bringing the total area covered by PSG to 2.58mha. This figure should be read against the background that an estimated 3.48m ha rank for such treatment. Fears that the new organization would imply an intolerable interference in the sanctity of private estates (to adapt the phrase used by Bazire and Gadant, 1991) have proved unfounded. Part of the reason for this is the important part that the local and central association of private forest owners play in the running of the national associations of CRPF. The 200 foresters of various grades have a heavy load: 1 for every 50,000 ha of private woodland, or 1 per 6800 owners. The total cost of CRPF operations, the bulk of which may be considered extension, amounted to 76.3m Fr.fr. in 1988 and 80.4m Fr.fr. in 1990. Together with a small input by Chambers of Agriculture the total cost of publicly financed extension is estimated at 85m Fr.fr.

Research

France has a large number of research institutes financed by the state and by wood-using industry. In 1990 total public expenditure on research bearing on trees and forests, excluding the (small) amount on arboriculture, is estimated to have been of the order of 120m Fr.fr. Private forests account for the following proportions of the country totals for: area 74%, increment 75%, (Fédération Nationale des Syndicats 1990). It is assumed that the proportion of research cost fairly chargeable to private forest owners is somewhat less than these ratios at 67%, implying research expenditure attributable to private forestry of 80m Fr.fr.

Summary of Public Expenditure in Support of Private Forestry

The estimated total public expenditure in 1990 is as follows (m Fr.fr.):

- Grants and loans 145
- Extension 85
- Research 80
- Total 310

Effects of Implementation Measures

To judge by the crude statistics of the areas planted and roads built with the aid of the Fonds Forestier National and drawing on the benefits of the land and income tax reliefs available, the reforestation aims of government have been successfully achieved, whatever view one may take of the modes of financing the activities or the the consequences for the distribution of income and wealth. Particularly important contributors to the programmes have been the extension services and the supporting research and development organizations, both private and public. The private sector has played a leading role in this. The whole purpose of the exercise is, however, the production of wood, and the greatest amount possible of sawlogs of good quality. Although a national forest inventory has been undertaken, the final step of comparing costs with the nature and quality of the resulting growing stock remains to be taken. Equally it is impossible on published evidence to say how far the public expenditure dispensed has succeeded in the timeous mobilization of the thinnings judged to be desirable or in decreased damage to fire-prone maquis and garrigue.

A valuable study leading to the characterization of the owners and their activities in the 46 Départements has been carried out by Normandin (1987). This might form the basis for a study of the effectiveness of the payments made and tax reliefs gained in the different geographical and ownership categories distinguished.

Conclusions on Outstanding Issues

Many of the arguments rehearsed in the policy debate in France and reflected in the panoply of measures developed are based on the assumption that the splitting of properties is bad and anything that achieves larger units of management is desirable. These points concern means more than ends. There appear to be no studies setting out the arguments for and against different scales of ownership nor of the real practicability of amalgamating management of scattered parcels of land. The contrast with the situation in Britain is remarkable since much policy in the latter recognizes the contribution of small woods, usually associated with small ownerships, to the appearance of the landscape and the protection of wildlife, with the only strongly expressed dissent coming from farmers who are typically not woodland owners.

The attention paid to conservation and, to a lesser extent, to landscape considerations is less than in other countries at a comparable stage of economic development. Implementation tools and procedures reflect this situation. Thus in the programme of the main public research institute, INRA, research into conservation is prominent by its absence (Birot and Lacaze, 1991). In addition, the concern of private owners to maintain their economic independence, a guiding principle in all countries, is particularly strongly expressed in France (see, for example, Peltereau-Villeneuve, 1992). The development of markets for wood needed to serve this ambition continues to be elusive.

The mix of laws, conditions, grants, loans and tax reliefs adopted in order to guide forestry in France is complex. It is presumably well understood by owners but it is, however, doubtful whether either its implications are fully known, or the aims of policy that the whole apparatus is intended to serve are lasting and therefore sound. The emphasis on the structural side is an outstanding example of this.

References

ANCRPF (Association National des Centres Régionaux de la Propriété Forestière) (1991) [*Annual Report 1990*] Paris.

Bazire, P. and Gadant, J. (1991) [*The forest in France.*] La documentation française, Paris.

Betolaud, Y. and Meo, J. (1978) [*The Wood Chain.*] Ministère de l'Agriculture, Paris.

Birot, Y. and Lacaze, J.-F. (1991) [Research on forestry and wood in INRA.] *La Forêt Privée* 202, 52–66.

Buttoud, G. (1979) [Private Forest Owners in France; Anatomy of a Pressure Group.] ENGREF, Nancy.

Buttoud, G. (1984) [Legislation on private forestry.] Paper to meeting of IUFRO S4.06.04, ETH Zurich. Typescript.

Buttoud, G. and Rochot, A. (1977) [The 30 year Relief of Restocking from Land Tax.] INRA, Laboratoire d'Economie Forestière, Nancy.

Clerc, F. (1980) [Forest policy: agents, institutions and development]. In: Société et forêts. *Révue Forestière Française*, Numéro Special.

de Jouvenel, B. (1978) [Towards the forest of the 21st century.] *Révue Forestière Française*, Numéro Special.

DERF (Direction de l'Espace Rural et de la Forêt) (Undated) [*Report on the Fonds Forestier National, 1989.*] Ministère de l'Agriculture et de la Forêt, Paris.

DERF (1989) [Modification of assistance by the Fonds Forestier National.] Circular reprinted in *La Forêt Privée* 33(192), 55–60.
DERF (1990) [*Guide to Forestry Grants.*] Ministère de l'Agriculture et de la Forêt, Paris.
Dubois, P. and Normandin, D. (1985) [*Forestry Cooperatives in Private Forestry.*] INRA, Laboratoire d'Economie Forestière, Nancy.
Duroure, R. (1982) [Proposals for a global forestry and timber policy.] *Révue Forestière Française*, Numéro Special.
Fédération Nationale des Syndicats de Propriétaires Forestiers Sylviculteurs and Association Nationale des Centres Régionaux de la Propriété Forestière (1990) [Summary.] (Statistics on private forestry). Paris.
FAO (1988) Forestry policies in Europe. *FAO Forestry Paper* 86, Rome.
Guillard, J. and Pardé, J. (1976) [France's wood resources and requirements.] *Révue Forestière Française*, Numéro Special.
Guillery, Ch. (1991) [From the forest to the wood using industry, a strategy for wood.] *Bulletin du conséil général du GREF* no. 31, 15–23.
Lagarde, M. (1991) [*Guide to the Legislation on Woods and Forests.*] ENGREF, Nancy.
Leclerc de Hautecloque, H. (1991) [The cadastral review.] *Forêts de France*, 340, 4–6.
Levêque, F. and Péguret, A. (1988) [*Forests and Wood-using Industry, structure and performance.*] Editions Economica, Paris.
Lowe, P. and Bodiguel, M. (eds) (1990) *Rural Studies in Britain and France*. Belhaven Press, London.
MacGregor, J.J. (1954) France's post-War forest policy. *Quarterly Journal of Forestry* 48(3), 167–178.
Martres, J.L. (1991) [*Report to the Forestry Association of the South-west.*] Bordeaux.
Mazodier, J. (1989) [Forest fires.] In: Stephan, J.M. (ed.), *La Forêt et le Bois* J.-P. Monza, France.
Ministère de l'Agriculture (1987) [Private forests.] *Service Central des Enquètes et Études Statistiques (SCEES)* Étude No. 268, Paris.
Ministère de l'Agriculture et de la Forêt. (1990) [The reform of the levy on forest products.] *La Forêt Privée* 33(196), 58–60.
Ministère de l'Agriculture et de la Forêt. (Undated) [*Plant and afterwards . . .*] Paris.
Mollière, C. and de Reure, G. (1988, with supplement 1989) [*Forestry Taxation Guide*] and [*Addendum, Wealth Tax.*] Centres d'Etude d'Economie et de Gestion de la Forêt Privée. Paris.
Normandin, D. (1987) [The management of private forests: structure and activities. An attempt to establish a typology for 46 départements.] *Révue Forestière Française* 39(5), 393–408.
Parrat, J. and Hocquet, A. (1976) [The modernisation of sawmills.] *Révue Forestière Française*, Numéro Special, pp. 130–145.
Peltereau-Villeneuve, C. (1991) [Towards a revolution in the forestry orthodoxy.] *Bulletin du conséil général du GREF* no. 31, 25–32.
Reed, J.L. (1954) *Forests of France*. Faber, London.
Wheare, J. (1990) State assistance to private forestry in Normandy. *Quarterly Journal of Forestry*. 84(1), 48–49.

8

BELGIUM

BACKGROUND

Management of Belgian forests was devolved to the country's three Regions – the Flemish (that is, Flanders), the Walloon (Wallonia) and Brussels – in 1980. Such devolution is of course common where there is a regional level of government, but at the same time Belgium took the less usual decision not to devolve agriculture, at any rate at that stage (it was taken in February 1993). The most obvious example in Europe of regional government controlling forestry is Germany, where, however, it is found in almost all cases alongside agriculture in the same Ministry. But Germany has three distinguishing features which make the situation in that country quite different from that in Belgium. The first is that the Länder follow the policy lead given by the Federal government, the second that the regard for forestry is high throughout Germany, the third that agriculture's institutional position and with it that of forestry in Germany is stable. The situation in Belgium is by contrast a developing one. The outcome for forest policy, its method of implementation and the operational result is not without interest in view of the lessons it offers on the administrative difficulties which arise with devolution for both government and private forestry. In Belgium the polarization of policies between Regions is marked and the pattern of government intervention surprising, if not, in some fields, perverse.

Salient geographical and physical points are as follows:

1. The total forest area for all Regions amounts to 603,000 ha, or 20% of the total land area of 3.08m ha, but forests are very unevenly distributed, covering 30% of the southern, Walloon Region and only 8% of the Flemish Region; the Brussels Region with only 2000 ha is not sufficiently important except very locally to warrant further consideration here.
2. The distribution of population also differs markedly:

(a) Flanders: population 5.74m, land area 1.35m ha, or 4.25 persons per ha;
(b) Wallonia: population 3.24m, land area 1.68m ha, or 1.93 persons per ha.

3. Private woodlands account for 55% of the country's total forest area and occur preponderantly in the Walloon Region; ownerships are generally larger in that Region. Areas are distributed thus:
   (a) private:
      (i) Wallonia 251,000 ha (51.5% of total for Region);
      (ii) Flanders 82,000 ha (72% of total for Region);
      (iii) total 333,000 ha.
   (b) State, communal, etc.
      (i) Wallonia 236,000 ha;
      (ii) Flanders 32,000 ha;
      (iii) total 270,000 ha.

There are major differences between Regions in the social characteristics of language, culture, and religion, as well as in economic structure. The following review concentrates attention on the Flemish position mainly because of the distinctive nature of the Region's policy aims. It also happens that policy has been further developed in that Region following the devolution of powers.

Current Legislation and Policy

Existing legislation, policy and implementation measures bearing on private forests are interesting from a comparative point of view. Although the situation is a developing one, it would be wrong to suggest that certain lines of policy are liable to significant change. The Flemish authorities issued a decree regulating forestry in 1990 (Ministère de la Communauté Flamande, 1990), a decade after formal regionalization and seven years after the formation of a new forestry administrative organization. As of 1992 there was no new forest law in Wallonia but the Conseil Supérieur des Forêts in the region had issued a discussion document (Conseil Superieur, 1991).

Meanwhile, arrangements for afforestation of setaside are the responsibility of the Ministry of Agriculture while it retained its central function. The resulting three-way split for forestry activities provided a good test of Regions' ability to achieve effective implementation of forestry operations in a relatively small country.

Flemish Region

Differences in the distribution of population and the degree of urbanization are influential in determining the emphasis of policy. In the Flemish Region there is a strong emphasis on multiple use throughout with the accent on conservation. The policy statement shows the following numbers of Articles devoted to each forestry function; 'economic', that is production of wood as a raw material, 2; social and educational (principally recreational and concerning access), 5; protective, 2; 'ecological' that is concerning conservation of native plants and animals, 4; scientific, including forest reserves, 9. Despite the relatively small amount of space given to wood and other marketable products, a fair degree of importance is attached to this function: the novel feature of the decree is the emphasis it gives to the other headings.

The aims are many; the following list records both general objectives and those specific to private woodlands:

1. Subject to nature conservation requirements, to promote the opening up of woodland for recreation and purposes of education.
2. To designate certain areas as protection forest.
3. To operate a rather conservative form of yield control, notably by restricting the cut in years following an unplanned and abnormally high level of cut undertaken for whatever reason.
4. To favour silvicultural methods which are closer to nature, specifically through:
   (a) conservation or reconstitution of the natural fauna and flora;
   (b) use of native species or strains suited to the site;
   (c) natural regeneration;
   (d) unevenaged stands or selection forest;
   (e) 'ecological equilibrium';
   (f) regulation of the use of herbicides.
5. To form forest (nature) reserves in perpetuity using management techniques outlined in **4** in order to conserve and develop woodland vegetation associations, stands and 'typical' forms of habitat.
6. To promote management according to the designated aims by making working plans compulsory for all private woods over 5 ha except in forest nature reserves where a nature conservation plan applies, and as part of this to require that private woodland owners conduct inventories of their properties every 10 years.
7. To encourage cooperative management in order the better to secure the desired benefits, such cooperative units being formed normally between private owners or occasionally between private and public sector forest owners.
8. To provide grant aid or compensation in support of these measures in private woodlands.

The flavour of the Flemish government's intention will be clear from this brief survey of the decree's main points. The degree of detail incorporated into the new law indicates the fundamental nature of the change of attitude and practice that is being demanded of private owners, their advisers and forest authority foresters. The occasion for the concern over the need for grouping of owners to achieve the government's aims is the fact that nearly three-quarters of woodland ownerships in East and West Flanders are below 1 ha (Ministerie van de Vlaamse Gemeenschap, undated).

Walloon Region

The discussion document issued by the forestry council of the Walloon Region (Conseil Supérieur Wallon des Forêts, undated) sets out four principal objectives:

1. To promote the increased productivity of the forest and the quality of its products.
2. To improve the profitability of private forests.
3. To increase the resistance of the forest to damaging agencies, both natural and man-made.
4. To maintain a judicious equilibrium between the material producing and other functions of the forest.

In respect of private forests in particular, the main aims are:

1. To discourage the break-up of properties.
2. To increase profitability.
3. To revise the *Code forestier*.
4. To carry out a forest inventory.
5. To control the genetic quality of new stock.

The Council makes a special point of recognizing that certain of these aims can only be achieved with the support of central government.

Administration

Unlike the organization prior to 1980, the forest services are englobed in larger administrations which for the most part deal with the rural environment. Thus in Flanders the forestry administration forms part of the division of nature conservation and development, itself within the directorate-general responsible for environment, nature and land management (Flemish forestry administration, undated). In Wallonia, where the scale of the operation is some four times larger, forestry constitutes a directorate general (FAO, 1988).

The Royal Forestry Society of Belgium combines the functions of representing the interests of forest owners, apart from the State, with those concerned with the forestry profession. It does not publish any financial results for forestry estates. The Flemish Forestry Society aims to appeal to forest-minded people; it is concerned with forest protection and extension as well as forestry practice which is close to nature.

IMPLEMENTATION

Whatever instruments regional governments devise to implement their favoured policy, their success has to take account of the fact that taxation remains a central government subject. In Belgium, as elsewhere in western Europe, forestry is not a particularly profitable activity and it has been found necessary to make special arrangements to meet this feature, for example by adjusting capital taxation to match the peculiarities of woodland management. Clearly the pressure for a more conservative form of management in Flanders will affect owners' finances, although grants will compensate for certain of the extra cash costs and losses. But it seems unlikely that, after taking these financial inducements into account, the tax regime adopted country-wide will prove suitable for a form of land use which will progressively depart from the conventional norms of silviculture practices aimed principally at wood production. Much will depend on the impact of the new objectives and practices adopted on capital values. It is relevant to note that taxation is also in the process of change as described below.

TAX

Although the system of taxation is determined nationally, one element, namely income taxation of agriculture and forestry is, as in many other countries, fixed locally (by commune).

There is no property or wealth tax.

Income tax

The basis is the cadastral revenue which ranges from 400 to 1600 Belgian francs (B.fr.) ha^{-1}. Varying tax rates are charged: the national (Belgium) element is 1.25%, the rate for the province varies between 3 and 20%, and for communes between 5 and 50%. Overall, an average tax rate of some 35% is believed to be representative (Terlinden, personal communication). This fraction is charged on 0.875 of the cadastral revenue, and the tax levied then calculated on the basis of the particular individual's (marginal) tax rate. Assuming average figures for the three elements, including the tax rate, a typical burden equals

$$1000 \times 0.35 \times 0.875 \times 0.50 = 153 \text{ B.fr. ha}^{-1}$$

Inheritance tax

Up to 1992, the basis of the tax on inherited woodland was the market value. The following calculation of a typical burden for the case of the property passing to a child is based on a capital value of 250,000 B.fr. ha $^{-1}$ (226,000 B.fr. in 1989, cf. Bary-Lenger and Stordeur, 1991), a tax rate for the slice of the estate bequeathed of 10%, and an incidence every 30 years. A tax burden of 25,000 B.fr. ha $^{-1}$ is implied. However, by the system of payment by annual instalments available in Belgium, the total amount due is two-thirds of this or 16,667 B.fr. a figure which may be translated into an annual equivalent assuming a multiplier of 0.0246 to yield 410 B.fr. ha $^{-1}$ year $^{-1}$.

From 1992, it is expected that the Senate will approve the proposal to give 75% relief on the market value on the two conditions that a simple working plan is prepared and that the owner maintains the area as forest for 30 years. Assuming unchanged capital values, the charge would thus be reduced to an equivalent annual sum of 103 B.fr. ha $^{-1}$.

Summary of tax burden for an average woodland owner

Adopting the 1992 regime as the basis of the assessment, the typical burden can thus be summarized as follows (B. fr. ha $^{-1}$ year $^{-1}$):

- Income tax 153
- Inheritance tax 103
- Total direct tax 256

GRANTS

Only small grants for the management of existing private woodland were given in the past in the Walloon Region; these being for first thinnings in coniferous stands. Unusually, publicly administered forests in both Regions were favoured with grant aid for a range of forestry purposes but private forests, with this one exception, were not. Since 1989, however, grants have been provided for a number of operations. In order to compare and contrast the grants adopted in the two Regions it is convenient to set these out operation by operation. Sources are, for Flanders, Ministère de la Communauté Flamande (1991) Vertriest (1991), and for Wallonia, Anon. (1991) and Terlinden (personal communication).

Planting

Flanders

Grants are payable for artificial or natural regeneration, on areas over 0.5 ha; payments are made in two instalments. Rates are to be reviewed annually. The owner has to agree to maintain the wood for 20 years. The basic grants are divided into five species groups as shown in Table 8.1.

Table 8.1. Basic rates of grant for planting or natural regeneration of stands, Flanders, 1991.

	B. fr. ha^{-1}
Class 1 sessile and pedunculate oaks, ash	100,000
Class 2 beech, cherry, hornbeam, *Acer* spp., *Tilia* spp., elms	80,000
Class 3 American oaks, sweet chestnut, alder, walnut, birch, willows, black poplar, abele, aspen, grey poplar, Scots pine	60,000
Class 4 Robinia, juniper, yew, Corsican pine, Douglas fir, Japanese and hybrid larch, grey alder, plantation poplar with broadleaved understorey	40,000
Class 5 plantation poplar	20,000

In addition where an understorey is introduced or is maintained which is composed of one or more of the following species, a premium of 20,000 B.fr. ha^{-1} may be paid: willow, alder, rowan, hazel, holly, alder buckthorn, guelder rose, spindle, dogwood, bird cherry.

Wallonia

Grants are payable for artificial or natural regeneration of broadleaved species only. The minimum area is 0.5 ha with an upper limit of 10 ha in any one year. Grants are paid in two instalments, at 6 months and 42 months after planting. The basis of payment is a proportion of the approved cost up to a defined maximum as set out in Table 8.2. The omission of conifers from this table is its most remarkable feature.

Table 8.2. Basic rates of grant for planting and natural regeneration of stands, Wallonia, 1991.

		B. fr. ha^{-1}
Sessile and pedunculate oaks, beech	70% of cost, up to	91,000
Black alder, Robinia	50% of cost up to	65,000
Birches, hornbeam, aspen, red oak, limes	35% of cost, up to	45,500
Sycamore, ash and cherry	30% of cost up to	39,000

In addition, a premium of 5% may be paid, each of which is cumulable, for the following: use of an approved Belgian provenance, natural regeneration, use in artificial regeneration of plants acclimatized for at least a year at an altitude not more than 150 m different from the site planted, plantations surrounding bird reserves, less favoured areas.

Weeding and cleaning

In Flanders only, grants for weeding and cleaning are available: weeding 5000 B.fr. ha^{-1}, weeding combined with crown forming pruning 8000 B.fr. ha^{-1}, cleaning 8000 B.fr. ha^{-1}.

First thinning

A grant of 5000 B.fr. ha^{-1} for first thinning of conifers is available in Wallonia only to a limit of 5 ha for any one owner in a year.

Grouping of woodlands

In Flanders, but not in Wallonia, grants are available to encourage the grouping of woodland properties for purposes of cooperative management. The rates are: 2000 B.fr. ha^{-1} for individual ownerships not larger than 5 ha and not more than 1 km apart, and 1500 B.fr. ha^{-1} for larger properties. A commitment has to be made to work to an officially approved plan established for a period of 20 years.

Provision of public access

Again in Flanders only, annual grants are provided up to a maximum of 1000 B.fr. ha^{-1} where hunting (shooting) is allowed by the landowner, and up to 2000 B.fr. where hunting is not carried out. Grants for path-making of 250 B.fr. ha^{-1} plus 20 B.fr. m^{-1}, are also payable in the former category, and 500 B.fr. ha^{-1} and 20 B.fr. m^{-1} in areas where there is no hunting.

Setaside

Neither Region has any specific policy for afforestation. It is noted in the Flemish decree that in addition to requiring the local agricultural officer's approval that of the mayor has also to be obtained before planting can go ahead. However, the Belgian Ministry of Agriculture has issued the terms under which the afforestation of setaside land is allowed country-wide (see Terlinden, 1988).

Loss of income is compensated by the Ministry of Agriculture, at rates varying with the part of the country, ranging from 8000 B.fr. ha^{-1} per year in the Ardennes up to 25,000 B.fr. on the higher quality silt soils of the lowlands and polders. These rates are reduced to 40% if the farmer decides to practise grazing on an extensive basis. Planting is not grant-aided. The extent of interest by farming interests in the whole scheme is clearly indicated by the fact that the area concerned in 1990–91 was 706 ha over the whole country. Afforestation within that total amounted to 76 ha, partly because of the scheme's odd feature of allowing only three genera to be planted, namely willow, poplar and alder. The point is made by the Forestry Society that the choice of species makes neither ecological nor economic sense (Terlinden, 1988; Ferdinand, 1990).

Extension

Each Forest Service has the usual subregional and district organization. Extension work is part of the remit of each Service. Until the advent of the new grant schemes, involvement by local Forest Service staff in the provision of advice was small. The Forestry Society feels that, given public funding, it is in the best position to carry out extension services because of its knowledge of owners and their circumstances.

Research

There are publicly financed research institutes at Gembloux in Wallonia, and in Flanders at the research institute for forestry and hunting with a section at Groenendal and the former institute of populiculture at Geraardbergen. These establishments are financed from central funds. It remains to be seen how satisfactorily the requirements of the Regions are met by this system. Staff input in 1991 was 65 person-years on all forest research, including that at universities. Allowing for a small amount of forest products research, total annual expenditure in 1991 is estimated at 200m B.fr. On an area basis private forests account for 55% of the total area in Belgium. The proportion of the cut accounted for by private growers is not known. It is assumed here that a reasonable allocation of the total research expenditure is 50%, implying an attribution of 100m B.fr. to private forestry.

Summary of Public Expenditure in Support of Private Forestry

Annual public expenditure in support of private forestry is thus estimated as follows:

Grants 17m B.Fr. (Flanders, 16m B.Fr.; Wallonia 1m B.Fr.)
Research 100m B.Fr.
Total 117m B.Fr.

Note that the amount given in grants to Wallonia excludes grant aid committed towards the storage of windblown material following the gales of 1990 of 75m B.fr. (Grollinger, personal communication).

Commentary on Implementation of Policy

Two features stand out from this review. The first is the resilience of a sector that has managed to continue to exist without major disposals of property to the state or other public body in the presence of amounts of grant aid that have been trifling until recently. It should be noted, however, that the fact that woodland values of the order of 160,000 B.fr. ha^{-1} in the high Ardennes, double those applying to much British woodland as measured by state forest sales, and as high as 500,000 B.fr ha^{-1} in Flanders, testify to the fact that owners find the possession of woods attractive. The social, amenity, hunting and other values are clearly substantial. In addition, the degree of control by the authorities is judged less onerous than in some neighbouring countries.

The second feature of note is the extent of the transformation of the financial aspect of woodland management likely to be affected by the schemes announced in 1991 by the two regional forest authorities. These have emerged following consultation: the resulting grant schemes display astonishing differences. The absence of a conifer planting grant in Wallonia reflects the profitable nature of spruce stands in that region, and it must be remembered that little afforestation, as opposed to restocking of well-grown conifer stands, occurs in the Region. The contrasting attitudes displayed by the different payments for such species as cherry and Robinia in the two schemes are especially piquant reminders of the fact that most grant schemes are essentially arbitrary. Although it is may be possible to rationalize such differences, curious attitudes are obviously represented by the favouring of particular species and such oddities and discrepancies must give owners cause for concern about the measures introduced in support of policy and even the continuity of policy itself.

Trends in Policy Issues

Because policy has so recently been promulgated in both Regions, it would be surprising if any new, unforeseen issue were emerging at the present time and this had not been noted and dealt with in the formulation of policy. Nevertheless it is possible to identify matters which appear to be in process of becoming more important with the passage of time. Broadly speaking, non-market outputs are likely to be given increasingly greater weight in

Flanders relative to Wallonia, forest industry development will remain and perhaps rise in importance in both Regions, while the more clearly economic factors of profit and income generation will be higher in importance in Wallonia than in Flanders. Although hunting is of overall lower importance in relation to other benefits of forestry, the fact that rents for hunting are higher in Flanders means that this activity is likely to remain important there, although compared with other recreation it may be ranked relatively more important in Wallonia.

It appears possible that the involvement of the authorities in forest management will become stronger in Flanders. Thus it has been reported (Jamblinne de Meux, 1991) that the nature conservation service had proposed, in the framework of an overall nature conservation plan, a certain designation of lands to be regarded as core areas for nature conservation and other areas where nature conservation should be promoted. The two categories happen to include all the forests of the Region.

The recent developments are of great significance for forestry. They are aimed at maintaining and improving the finances of forestry while enhancing non-market benefits, to a fair degree through the favouring of forestry practices that are close to nature. Private owners have little option than to accede to these changes.

References

Anon. (1991) [Grants for the regeneration of broadleaves in the Walloon Region.] *Silva Belgica* 98(4), 48–51.

Bary-Lenger, A. and Stordeur, P. (1991) [Private forests and inheritance tax charges in the Walloon Region from 1950 to 1989.] *Silva Belgica* 98(3), 7–18.

Conseil Supérieur Wallon des Forêts (1991) [Statement of forest policy.] Typescript.

FAO (1988) Forestry policies in Europe. *FAO Forestry Paper* no. 86, Rome.

Ferdinand, C. (1990) [The effect of setaside on the planting of farmland.] *Silva Belgica* 97(3), 29–31.

Flemish Forestry Administration (Undated) The Flemish forest service. Typescript Brussels.

Jamblinne de Meux, A. (1991) Editorial. *Silva Belgica* 98(4), 4–5.

Ministerie van de Vlaamse Gemeenschap (Undated) [*Forestry Decree: a New Forest Law for Flanders.*] Brussels.

Ministère de la Communauté Flamande (1990) [Forestry decree, 13 June 1990.] *Belgisch Staatsblad*, 28 Sept. 1990.

Ministère de la Communauté Flamande (1991) [Regulations concerning grants to private woodland owners and the agreements for cooperatives.] *Belgisch Staatsblad* 16 September.

Terlinden, M. (1988) [The point about the afforestation of farmland.] *Bulletin Societé Royale Belge* 95(5), 18–20.

Vertriest, I. (1991) [Executive decisions of the Forest Decree (Flanders Region).] *Silva Belgica* 98(3), 39–47.

9

THE NETHERLANDS

BACKGROUND

Six facts may be noted as highly relevant to the consideration of forestry, and particularly private forestry, in The Netherlands.

1. The population density is the highest of any European country at 4.4 ha^{-1}, a figure which may be compared with 3.2 for Belgium, 2.4 for Britain, 2.3 for Germany and 1.0 for France.
2. Although woodlands occupy some 10% of the land area of the country, the area of forest land per head of the population is the smallest in the developed world at 0.02 ha.
3. The potential for woodland to add another dimension to countryside recreation is great.
4. Virtually nothing remains of the natural woodland that once covered the country and was largely removed by the 16th century, and old woods are few, only 16% of the forest being over 80 years.
5. A substantial area, 37,000 ha, of woodland is owned by private conservation bodies.
6. In contrast to the position in almost every other country, there has been a steady decline in the area of private individuals' forest holdings from 65% in 1938–1942 to 41% in 1980–1983.

Taken together with the nature of the Dutch character, the first two points combine to produce an attitude to land use planning which is probably the most careful and deliberate of any European country. A particular expression of this that bears on woodland is the desire to provide wild environment both for general recreation and as habitat for wildlife and its enjoyment by nature lovers. Although the word Holland is derived from 'holz-land', the sparseness of woodland relative to total land area is marked. And although, for some, forests represent a form of visual pollution because of the barrier they present to the eye in a flat landscape, the pressure for more woodland and access to it is great. The importance of community interest in woodland ownership is demonstrated by the area statistics in Table 9.1.

It is hardly surprising that a main feature of environmental organization in The Netherlands is the concern with rural planning (Ministry of Agriculture, undated, illustrates the variety and complexity of rural development projects ranging from improvement of the efficiency of drainage to the consolidation of holdings). Such land use planning involves central government in some detail as well as the province and the local municipality. The demands for specialist uses of woodland as well as for what may be broadly termed 'social' uses have increasingly put pressure on private individuals' woodland holdings. This and the low profitability of forestry have led to the decline in the area of forest owned by individuals. Although private woodland remains the largest single wooded area within the total for forest land, and easily exceeds the area of 'forestry' land in the State forest service's hands, its decline must be regarded as highly significant. The nature reserve categories together represent 15% of all forest land.

Within the area which is stocked with trees, a notable feature is the youthful age distribution. With a high proportion of stands established post-1930, it is therefore not surprising to find that annual increment at 2.52m m^3 (average for 1982–1989, Busink, personal communication) substantially exceeds cut at 1.3m m^3.

Table 9.1. Distribution of forest area by ownership, 1980–1983.

	Area (ha)
State	
For forestry use	62,291
Nature reserves	11,851
Other land	30,181
Subtotal	104,323
Local authorities	
Provincial	2,472
Municipalities	50,258
Other public authorities	2,354
Subtotal	55,084
Private	
Nature preservation societies	37,486
Other private owners	136,385
Subtotal	173,871
Grand total (including 749 unidentified)	334,026

Source: Centraal Bureau voor de Statistiek (1985).

Financial Position of Private Forestry

Table 9.2 summarizes data on current revenue, including grants, and expenditure for a sample of estates over 50 ha, a class that accounts for 75% of private forests. The data come from an identical sample.

The contribution of timber sales to the total receipts of estate woodlands typically lies between 40 and 50%. Standing timber prices are reasonably high by British standards with values in October 1991 across all tree sizes varying from gld. 35 to 55 m^{-3} for conifers and from gld. 150 to 250 (van Woudenberg, personal communication). No published calculation is available to show the value of change in growing stock in private woodlands. It seems likely that increase in the market value of this element of capital easily offsets the cash deficits averaging some gld. 25 ha^{-1} over the last 10 years.

The state imposes two major requirements on private woodland owners. One, the payment of substantial water board rates, is an unavoidable feature of the occupation of land in The Netherlands. The other group of obligations revolves around planning activities. These include the need to gain the agreement of local authorities in relation to certain planning functions in the rural environment. In forest outside the environs of towns, the Forest Law applies and there are no difficulties concerned with normal management. But the consultations required in the preparation of municipal land use plans are very time consuming for landowners. Equally the administration of forest grants and provision for allowing relief from inheritance and other taxes on forest land are time consuming. The rewards are valuable in terms of reduction in or relief from tax, and various grants are available for planting, later tending, and the preparation of management plans.

Table 9.2. Financial results for private woodlands.

	Guilders (gld)[a] ha^{-1}									
	1981	'82	'83	'84	'85	'86	'87	'88	'89	'90
Total costs[b]	497	504	465	461	575	424	432	443	378	386
Total revenue[c]	426	473	421	426	553	445	411	393	378	406
Net revenue	−71	−31	−44	−35	−22	+21	−21	−50	0	+20

[a] gld. is used for clarity in place of f., the Dutch convention.
[b] Including water board rates.
[c] Including grants.
Source: Bedrijftsuitkomsten in de Nederlands particuliere bosbouw over 1989, 1990 (updated). Landbouw Economisch Instituut, The Hague.

LEGISLATION

Because the law stands, as in Britain, as the framework within which policy is adjusted to the needs of the time, only brief reference needs to be made to salient laws bearing on forestry. The critical laws are:

1. The Forest Law 1961 (effective 1962). This requires forest to remain forest, with restocking required after final felling and usually on the same site. An owner wishing to fell simply informs the forest service of his intention. Although the state can prohibit felling for a period of 5 years, the power is only used exceptionally.
2. As elsewhere, the basic forest law is concerned with the conservation of a tree stock but pays little attention to the nature of the resource that it safeguards. The Natuurschoonwet 1928 (often translated as the Nature Scenery Act, which is effectively concerned with landscape protection and here referred to simply as the Landscape Act in order to distinguish it from the Nature Protection Act) provides important tax incentives to landowners who undertake works to enhance the natural beauty of their estates and allow public access. The principal requirements are that the estate should have not less than 30% under woods, that the total estate should be not less than 5 ha and that consistent management be applied. A special regime which gives relief from capital taxes applies when paths are open at the rate of at least 50 m ha^{-1} in woodland and 25 m ha^{-1} on farmland.
3. The Nature Protection Act 1967 provides for the registration of sites of wildlife conservation value. Rigorous controls are written into the management plans for these sites agreed with the owners.
4. The Land Use Act directs that municipalities make land use plans. The framework is determined at provincial level, and the plan developed in detail at the local level by the municipality. The plans, which are revised every 10 years and are more detailed than in Britain, may of course have important implications for estate management because of the designations they incorporate.

POLICY

The government statement of forest policy in FAO (1988) is remarkable not for the information it provides so much as the attitude to forests that it reveals. Thus: '(the) main objective of the Government's forestry and silviculture policy is: the promotion of conditions and circumstances within the framework of total government policy such that the forested area in The Netherlands, in terms of size and quality, corresponds as nearly as possible to the wishes of society regarding the use of forests, now and in the future, at a cost level acceptable to society.' In what is a masterpiece of drafting covering the concerns of almost all other conceivable missions of government, the striking feature is the emphasis on society. This is more reminiscent of a language and philosophy of western Europe one or two decades earlier than of the 'market-oriented' emphasis of more recent years, although the last phrase may be interpreted to mean that effort must be devoted to gaining as large a revenue from saleable items as practicable without prejudicing the provision of non-timber benefits. Overall, however, the implication of the statement is that society, in practice most obviously and effectively through the municipalities, has an important role to play in calling the tune, subject in the outcome to checking the cost to the taxpayer.

A major policy decision was set out in the Long Term Programme (Ministerie van Landbouw, 1986) which has the following central aims:

1. 82% of the forest has to be multifunctional (it should also be noted that the remaining 18% is almost entirely devoted to preservation of existing trees and plants).
2. The forest is to have a more varied composition with longer rotations.
3. The forest area must be extended by 30,000 to 35,000 ha.
4. Removals must be increased to 2m m^3 p.a.

The final aim is to cover 25% of the need for forest products by domestic roundwood production, in order to diminish the dependence of Dutch wood processing industry on imports of raw materials. The target date for the afforestation and the doubling of cut is the middle of the next century. The clarity of the two statements is exemplary.

One element of the planting of new forest requires the creation of 10,000 ha of dual-purpose timber-producing and recreation forests in the 'Ringtown' scheme in the west of the country. The acquisition, establishment and management of these is to be a state forest enterprise responsibility (Swellengrebel, 1989, 1991).

Administration

Three main authorities are responsible for forestry: the Ministry of Agriculture, Nature Management and Fisheries, which incorporates the state forest service (Staatsbosbeheer); the Ministry of Economic Affairs, which is responsible for wood supply and wood-using industry; and the municipalities. The state forest enterprise was separated from the forest authority in 1988.

The Ministry of Agriculture ensures that national policy is implemented appropriately at provincial and municipality levels. The intention is thus that the government should work through provinces and municipalities with their land-use plans. These bodies may also, if they so decide, make

Landscape Structure Plans, one element of which may be consolidation of holdings (Bakker, 1990), and Nature Protection Plans. Although land with special use or 'geographic' or other constraints, such as woodland, is not subject to reallocation under the terms of the Land Development Act, 1985 in the same way that agricultural land is (Grossman and Brussaard, 1989), it is important to note the extent of these powers and the degree of control they imply over a landowner's activity.

Arrangements for afforestation with fast-growing species are the responsibility of the Ministry of Economic Affairs, an arrangement rationalized on the grounds that the interest in extensive planting of such relatively fertile land lies primarily in wood production. The sole species favoured are willow, poplar and Norway spruce. Special provision has been made so that there is no obligation to restock such woodlands. Environmental groups and many local authorities have expressed disquiet over the scheme. In addition the Ministry of Agriculture supports the afforestation of setaside land.

A major concern in The Netherlands is the satisfaction of the demand for country recreation. There is a 'Structural Scheme for open air recreation'. It has been estimated that 185 million visits are paid to forests annually: private forest owners play an important part in catering for this use through the workings of the Landscape Act.

The industry board for forestry and timber growing (Bosschap) brings together employers' and employees' organizations and serves as the permanent advisory body to the government. Owners of more than 5 ha of forest are required to register their woodlands with the board and are charged a levy (Linnard, 1988). The board is empowered to issue regulations relating to the protection of forests and working conditions of labour. The Netherlands Association of Forest Owners (Nederlandse Vereniging van Boseigenaren) is open to all owners apart from the State. Its members account for some 120,000 ha of woodland or 52% of the non-state total.

Tax

Income tax

There is no income tax payable in respect of woodland.

Water board tax

This tax, which varies from board to board, constitutes a considerable burden for forestry. For the sample of private woodlands over 50 ha, it averaged, with little variation over the period 1981–1989, gld. 25 ha^{-1}. The charge is adjusted periodically.

Property tax

Municipalities are empowered to levy this tax; forests are exempt.

Wealth tax

Forests are subject to this tax, which is levied at a flat rate of 0.8% on assets over gld. 250,000, but are relieved where the woodland has satisfied the terms of the Landscape Act: (i) by a reduction of 50% in the value of woodland including timber; and (ii) wholly where the property is open to the public.

Assuming an average value for woodland of gld. 10,000 ha^{-1}, and an average rate of tax over the whole of a person's net wealth of 0.6%, the tax would amount to 0.6% × 0.5 × 10,000 or gld. 30 in case (i).

Inheritance tax

Similar provisions to those of wealth tax apply, with the principal difference being that the beneficiary is only granted relief if the estate of which the woodland forms part is managed as required by the terms of the Landscape Act for 25 years. Assuming an average rate of tax of 15% (*Fiscaal memo*, 1991) and a factor for the annual equivalent of a lump sum payment of 0.03, the annual tax burden where the Landscape Act is not invoked is 0.15 × 0.5 × 0.03 × 10,000 or gld. 23 ha^{-1}.

Summary of direct taxation charges

Average amounts of tax are estimated as follows for the owner of an average forest holding (expressed as gld. ha^{-1} $year^{-1}$):

Income tax	zero
Water board rate	25
Wealth tax	
if Landscape Act applies	zero
otherwise	40
Inheritance tax	
if Landscape Act applies	zero
otherwise	23
Total direct tax	25–88

Grants

Forestry grants administered by the Ministry of Agriculture are revised intermittently, most recently in 1990 (Anon., 1991). They are paid at the rates set

Table 9.3. Grant rates and stocking densities for various species.

	gld. per plant	Stems ha^{-1}	Range of grant (gld.)
Group I			
Grand fir, hemlock	1.45	2500	3625
Douglas fir	1.4	2500–3500	3500–4900
Alders, birches, bird cherry, gean, *Robinia*, large leafed lime, abele, grey poplar, aspen, sallow, rowan	1.15	2500–4500	2875–5175
Weymouth pine, maritime pine	1.15	2500	2875
Silver, fir, Norway and Sitka spruce	1.1	2500	2750
Larches	1.1	2500–3500	2750–3850
Red oak, sycamore, beech, ash, hornbeam, small leafed lime, sessile and pedunculate oaks	1.05	4500–6500	4725–6825
Scots, Corsican and Austrian pines	0.95	2500–4500	2375–4275
Group II			
Populus euramericana, *P. interamericana*, *P. nigra*, *P. trichocarpa*, *Salix alba*	7.0	200–625	1400–4375
Group III shrubs[a]			
Includes, elders, broom, sea buckthorn, hawthorn, hazel, medlar, guelder rose, wayfaring tree, dog rose, privet, snowy mespil, willows	1.4	4500–6500	6300–9100

[a] Not normally applicable to forestry. Other grants, for Groups IV (avenue trees) and V (hedge species), are not listed here.

out in Table 9.3 which are based on the concept of meeting ca.75% of actual costs. Payment is conditional on there being a current management plan; this is revised every 5 years. As will be seen, there are highly specific rates of grant within each major operation.

Restocking and normal[1] afforestation

For each species a grant rate per plant is set together with a range of densities per hectare on which grants are payable (Table 9.3). The subdivision by species is very detailed.

Stand improvement

Two operations are assisted (Table 9.4). There is no grant for drainage improvement.

Table 9.4. Stand improvement.

	Rate per tree	Number	Rate (gld. ha^{-1})
Marking of final crop trees		60–120	210
Pruning			
Poplars	max. 3.5	400	max. 1400
Douglas fir, larches		min. 60	650
Scots pine		min. 60	350

Recreation provision

The grants are highly specialized in this area and only those for some principal facilities are listed here.

- play area
  gld. 0.45 m^{-2} on sandy ground
  gld. 2.1 m^{-2} on non-sand with drainage
- car park at main entrance to facility
  gld. 184 per place on sand
  gld. 326 on non-sand
- car park for entry route
  gld. 22 m^{-2} half hardened,

[1] Normal: i.e. not fast-growing species or setaside.

gld. 35 m^{-2} hardened on sand, gld 51 m^{-2} elsewhere
- bridge on entry route gld. 686 m^{-2} simple span, gld. 868 m^{-2} with piles
- paths: very many specifications not listed here.

Management

The rate payable when a management plan is in operation is gld. 65 ha^{-1} per year, or gld. 90 where open access to woodland is allowed by the owner. This constitutes the largest single grant in terms of total value.

Afforestation with fast-growing species

A premium of gld. 3000 ha^{-1} in addition to the normal grant is paid for planting of fast-growing species such as poplar, willow and Douglas fir where these may be expected to achieve specified minimum production rates, e.g. poplars not to be cut before 15 years or 200 m^3 ha^{-1} attained, with no crop life extending beyond 40 years.

Setaside planting

In addition to the normal planting grant, the planting of fast-growing species on setaside land entitles the owner to an annual grant of gld. 1500 ha^{-1} for the first 50% of the agreed area of setaside and gld. 1300 ha^{-1} for the remainder.

Total grant aid

Total forestry grant aid of all kinds amounted to gld. 39 m in 1990–1991 (Director General, State Forest Service, 1991). A substantial proportion of this total is, however, paid to municipalities and private nature protection bodies and it is estimated that the element due to private individual owners was gld. 33m.

EXTENSION

Extension was introduced as a specific activity in 1988 following separation of the forest enterprise from the forest authority. Figures on extension expenditure in respect of private woodland owners (individuals) are not specifically recorded.

Research

Some 145 professional and technical staff were employed on forest research by central government in 1991 (Hinssen, personal communication). The number employed by local authorities is not known but estimated not to exceed 10. The total cost of the services provided is estimated at gld. 18m in 1991. Allocating expenditure by reference to forest area would suggest a proportion of some 41% attributable to private owners, and by reference to planting (new planting plus restocking) in the 1980s 32%. For working purposes a figure of 33% is adopted indicating a value of research allocable to private forestry of gld. 6m per year.

Total Public Expenditure in Support of Private Forestry

In sum, amounts paid to private forest owners or expended on their behalf in 1991 were: grants gld. 33m; research gld. 6m.

Commentary on Success of Implementation Measures

The fact that the area of private individually owned forest has declined so substantially since the mid-1960s suggests that the costs of carrying on forestry, even with the tax reliefs and grants offered, are seen by many owners to be too high. It is notable that the government's statement of policy makes no direct reference to ownership. It might be assumed that the emphasis in the general statement on 'society' implies indifference towards the continuation of private (individual) ownership of forests. Against this, it is interesting to note that it was intended that from 1992 municipalities should no longer rank for woodland grants.

It is clear that the fact that many interests, private and public, claim that woodland is necessary for them to achieve their aims puts a heavy burden on a very limited resource. Cash injections into private forestry amount to over gld. 30 m and tax reliefs at a minimum valuation to a similar amount, that is a total of some gld. 60 m. This is equivalent to gld. 440 ha^{-1} per year. But it is by no means clear that the country's goals for the many functions can be met by the current system of financial inducements. It is note worthy that the government has established a commission to review private forestry grants. Of equal importance, the government has indicated that it is concerned with the cost-effectiveness of its policies. The 1992 evaluation exercise is, however, more explanatory than analytical.

Outstanding Issues

Although the aims of land use planning are clear and the technical aids available are powerful, there appear to be two major difficulties in its effective implementation. The first is that the relation between central and local government (largely though not exclusively the municipalities) has not been stabilized in the forestry area. Secondly, the attitude of government towards the role of private owners in meeting goals which are heavily oriented towards non-market outputs is ambivalent. Against this background the tension introduced by schemes such as that for afforestation of setaside land is obvious.

Bureaucracy is a major factor in the implementation of any finely divided system of grants as in The Netherlands. In addition to the time required of owners or their agents, substantial administrative costs are incurred by the authorities in pursuing the achievement of public goals through the support of private forestry.

References

Anon. (1991) [Grants for forestry.] *Bosbouwvoorlichting* (1), 15–18.

Bakker, A.J.J. (1990) [Natural landscape: still not all questions resolved.] *De Land Eigenaar* 36(9), 8–9.

Centraal Bureau voor de Statistiek (1985) *Nederlands bos statistik*. The Hague.

Director General, State Forest Service (1991) Address to Czechoslovak delegation. Typescript.

FAO (1988) Forestry policies in Europe. *FAO Forestry Paper* no. 86. FAO, Rome.

Fiscaal memo. 1 (1991) Kluwer, Deventen.

Grossmann, M.R. and Brussaard, W. (1989) Reallocation of agricultural land under the Land Development Law in the Netherlands. *Studies on Physical Planning* 3, Wageningen Agricultural University, Wageningen.

Landbouw Economisch Instituut (1990) [*Management Results for Netherlands' Private Forests During 1989*.] The Hague.

Linnard, W. (trs.) (1988) *Forestry in the EEC: Kingdom of The Netherlands*. Brussels.

Ministerie van Landbouw, Natuurbeheer en Visserij (1986) [Government decision concerning the forest policy plan.] *Tweede Kamer*, 1985–1986, 18630, nos. 5 and 6, Den Haag.

Ministry of Agriculture, Nature Management and Fisheries (Undated) *Rural Development: a Dutch Experience*. The Hague.

Swellengrebel, E.J.G. (1989) [New forests in the ringtown region.] *Nederlands Bosbouw Tijdschrift*, 240–246.

Swellengrebel, E.J.G. (1991) Afforestation experiences from the ringtown Holland; opportunities and constraints in densely populated areas. Paper to EC workshop on the afforestation of set-aside land. Brussels.

10

GERMANY

INTRODUCTION

The unification of Germany in October 1990 has meant that a new situation is in process of development. Forests cover some 7.4m ha of the former Federal Republic's land area of 24.9m ha and for 3.0m ha of the five eastern Länder, that is in total 10.4m ha or 29% of the country (Gallus, 1991). Private forests covering some 3.5m ha account for 45% of the total forest area of the former FRG, and in 1936 amounted to 52% of the forests of the five eastern Länder (Anon., 1991a). At the time of writing, privatization of former East German forests which were nationalized by the then USSR government between 1945 and 1949 had not started owing to uncertainties over the legal basis of this change of ownership. Privatization of forests nationalized after 1950 had begun. Because the purpose of this survey is to identify the interactions between policy and the private sector, the following discussion emphasizes the position in the territory of the former Federal Republic, here termed, for convenience, West Germany.

Several qualitative features may be noted as relevant to the consideration of German forestry:

1. Through the creation of scientific schools as early as the latter part of the 18th century, German forestry traditions are well established and have long been respected by both those within and outside the profession.
2. The well-being of forests is perhaps of more concern to the country as a whole than is true of any other European nation outside Scandinavia.
3. There is a wide range of site conditions including, in the south of the former east Germany, areas subject to some of the most intense air pollution of any region in the western world.
4. Forests are intensively managed along traditional lines (close spacing, light thinnings, emphasis on native species and the production of high

quality sawlogs) and their income-generating capacity is important to all classes of owner.

5. Environmental groups exert a heavy pressure on foresters to adopt 'near-to-nature' forestry, or in the extreme to abandon forest management altogether.
6. Hunting is an important use of forest and may prejudice the application of cheap and effective silviculture.

Gallus (1991) notes that the bulk of the forest estate is coniferous, 67% in the west and 72% in the east. The growing stock recorded (ECE/FAO, 1992) for the western part of 290 m^3 ha^{-1} is high by any standard. Wood production remains an important goal of forestry in Germany. Total cut in West Germany in 1989 was 31m m^3 under bark, which, together with fibre derived from waste paper, meant that the country satisfied some 60% of its wood needs from domestic sources[1]. This result has been achieved through the combination of a competitive forest industry, parallelling the log prices in such countries as Finland and Sweden, and control of harvesting costs, again, as in Scandinavia, in the face of impressive rises in wage rates. The desire on the part of owners and their forest managers to maintain, and usually to maximize, the production of high quality, valuable wood is, however, at the root of much of the discontent expressed in recent decades with the objects of management adopted in German forests. These reflect the contrasting and growing dissatisfaction flowing from divergences between the aims of owners and the goals of others, often arrayed in powerful environmental pressure groups.

Historical Background

To many foresters, Germany appears to display a strong forest tradition, strict in its methods and protocols and of a uniformly high standard. The history of the 19th century shows how varied has been the course of true forestry in that country. After the French Revolution and the new order subsequently established in Europe, a number of changes occurred. These were more marked in areas where French influence was marked, as in the Rhineland, where the impossibility of undoing some of the things Napoleon had done determined the course of events (Thomson, 1966). But mostly the landed aristocracy held on to their estates. There was, however, a marked relaxation of the control operated in Germany under the absolutist regimes of the previous two centuries (Troup, 1938). State Diets were restored, in

[1] The basis adopted in Germany: the measurement of self-sufficiency in physical or money terms is always a difficult process.

some regions the hereditary domains of the princes were converted into State forests and, as Greeley (1953) argues, as a result of the ideas of Adam Smith, restrictions on the use of remaining private woodlands were relaxed. This led in the case of the northern lands, where the owners were more powerful, to some uncontrolled deforestation (Klose, 1985) but also, because of the pressure of industry, to the afforestation with or conversion to plantations of conifers, especially Norway spruce. Zon and Sparhawk (1923) observe that imports of wood overtook exports after 1863: this acted as a stimulus to more intensive practice. In line with the more liberal political conditions, there was even some sale of state forests and state control of markets and prices disappeared.

Contemporary with the ebbs and flows in the pattern of ownership, acceptance of the sustained yield concept in forestry was growing in the second half of the 19th century. Working plans for public forests divided larger forests into small economic units, this tradition being one that is carried down to the present day as observed in State forest practice in the 1990s. Thus, in Lower Saxony, the division into territorial charges is to districts averaging 5000 ha, and within them to ranges or beats under the subprofessional forester grade averaging 900 ha (Ripken, 1991). In the last century there was a pervasive attitude towards forest land as the basis of family security; indeed in 1938 about 27% of the area of private forest was entailed (Troup, 1938). The implication was that the owner acted as steward of the property with the government controlling the management even within states which otherwise made private forests the subject of next to no control. After Bismarck, the southern states partly restored controls over private forestry which, as Greeley remarks, had been swept away in the preceding wave of industrialism, but this control was still only important according to Zon and Sparhawk (1923) in Baden and Württemberg with control of felling and obligatory restocking and intervention by the state in the case of poor management.

The experience in 1923 of dramatic hyperinflation had important effects. Crowther (1940) records that following the great inflation of the Weimar republic, there was an urgent demand for a 'good' money and a new currency, the Rentenmark, was produced which was said to be backed up by the land of the country. Crowther points out that there was no method by which the holder of a Rentenmark note could possess himself of the land that was supposed to be behind his note. But the elaborate bluff worked, and so strong was the belief of the people that any money which either consisted of, or represented, a valuable commodity was an efficient money, that the Rentenmark was accepted. The experience of 1923 left an imprint with reverberations carrying well into post-war West German history (Fulbrook, 1990). Foresters often claim that one of these was the even greater emphasis on heavy growing stocks in both public and private forests.

The more recent history has included the period of Nazi rule which was marked by a strong central control, including creation of a cartel, control of trade, and central authority decision on the cut of every individual owner. A draft law of 1942, which in fact never came into force, had as its main concern the use of the German forest mainly as a source of raw materials. Rozsnyay and Schultze (1978) have argued that the draft was of special importance because of the influence it had on forestry legislation at the Federal and Land level after the Second World War, for example in the emphasis this gives to forest service supervision of private forestry and to the protection and recreation functions of forests. In 1945, the Occupying Powers decided to restore the pre-Nazi position whereby the administration of agriculture and forestry was a regional (Land) responsibility, this being confirmed in the founding of the Federal Republic in 1949.

Traditions play an important role in forestry. An example is provided by the recent measures, such as those taken in Bavaria, to support small family farms about half of which were part-time, and many of which have a small forest associated with them (Eisenmann, 1983). Farm forests averaging about 5 ha constitute nearly half of the former FRG private forest area. Roughly 44% of farms have woodland (Giesen, 1991a). The owners of such properties clearly have an important political influence. The historical tradition also applies, and with even more force, to the distinction between public and private interest and the concept of the need for public control, aspects that underlie the intense discussion on the damage caused to German forests by air pollution, and the more pervasive and demanding issue of adjustment of forest management to meet multifunctional demands. Yet foresters display alarm at the expression of public interest which has become so marked a feature of the past few decades. It seems reasonable to ask how far lessons have been learned of the way in which attitudes can change and, in the end, institutions have to alter.

The Economic Position of Forestry

The range of sizes of forest property in private hands is very wide as shown in Table 10.1. In particular the first two size classes and especially the second indicate the importance of farm forests, many of the smaller holdings being associated with farms.

A feature of interest to the present study concerns the ability to assess the financial performance of State, communal and private forests. Very long data series, covering centuries, for net annual cash income are only available for state forests, useful reviews of which are provided by Brandl (1987) and Ott (1987). In addition, though for rather more recent times only, Germany is one of the few countries in which data are available which allow comparison across ownerships to be made.

Table 10.1. Size-class distribution of German forests (Western Germany).

	Size-class (ha)					
	<1	1–50	50–200	200–500	500–1000	1000+
Number (rounded)	110,000	320,000	4200	800	240	160
Area (thousand ha)	49	1612	378	246	167	428
Area (%)	2	56	13	8	6	15

Source: Anon. (1991b: [Enterprises with forest by ownership and size.], Table 94).

For the purposes of illustration, financial results for one Land, Baden-Württemberg, are set out in Table 10.2.

Although the figures fluctuate they indicate clearly the contrasting performance of the different sectors. The dominant features are first that farm forests (the sample uses the range 5–200 ha) show higher surpluses than private forests of over 200 ha, and that communal forests produce better results than State forests and, secondly, the negative results for the state forests. These findings represent a position which appears to be repeated widely in Germany and for forests which are, by and large, going concerns (see, e.g. Behrndt, 1989; Anon., 1991a) though with fluctuations in growing stock levels caused by storms, economic conditions, etc. The interpretation of the differences among ownership categories in levels of net annual income is more difficult. It is reasonable to assume that both physical attributes and management objectives influence the marked differences observed.

Behrndt (1989) has tabulated time series from 1951 for net annual income for the three categories of ownership for west Germany. If the results are deflated by the implicit GDP deflator, the pattern of net income in private woodlands shown in Table 10.3 is found (for convenience the data have been summarized by quinquennia). The dramatic decline after 1955 is remarkable and the subsequent stabilization, followed by recovery in the 1980s, noteworthy and in contrast to the state forest results which remain negative though not as low in real terms as during the late 1970s. The different reactions to market forces have almost certainly had a good deal to do with these differences in performance. In addition there are no doubt variations among ownership categories in the level of the cut relative to increment. Finally it is noteworthy that at present the average net income of the order of 100 DM ha^{-1} per year represents about 1% of the value of the

Table 10.2. Net annual income[a] of different categories of ownership.

	DM ha^{-1}								
	1983	'84	'85	'86	'87	'88	'89	mean[b] 83–89	'90[c]
Farm woods	192	368	94	133	91	31	228	162	876
Other private	165	133	77	46	−22	84	153	91	721
Communal	131	142	49	63	17	−3	61	66	465
State	34	101	−123	16	−97	32	134	14	668

[a] 'Reinertrag', in principle (*sic*) excluding value of owner's own labour, but ignoring changes in capital stock.
[b] Average of money sums unadjusted for inflation.
[c] Shown separately because this year includes the effect of large volumes of windblow.
Source: Anon. (1991c).

Table 10.3. Net annual income for private forests (former FRG).

Quinquennium	Net annual income (DM (1980) ha^{-1})
1954–1958	323
1959–1963	93.2
1964–1968	19.2
1969–1973	17.2
1974–1978	39.2
1979–1983	59.4
1984–1988	58.0

Source: Behrndt (1989, Tables 39 and 62) and author's calculations of real terms values.

growing stock if this is conservatively valued at 12,000 DM ha^{-1} (60 DM m^{-1} on a volume of 200 m^3 ha^{-1} for the average of private woodlands).

Legislation and Policy

As in many other continental countries, policy is expressed in laws, often of a most detailed kind. Indeed the dependence on a legal basis is regarded as paramount. Thus Niesslein (1991), commenting on the Federal Constitutional Court's 1990 decision apparently overturning the objectives of the Forest Act as described below, observes in a rather magisterial manner that policies may set new goals for management but as long as they are not conditional or otherwise limited and the relevant written laws are not amended, such statements are irrelevant. Secondly, although forestry is a 'Land' subject, the Federal government in effect sets the agenda within which Land governments act. Thus the Federal Forest Act of 1975 (Federal Ministry, 1984) states that the Regulations of the chapter on preservation of forests constitute a framework for legislation at state level. Examples are provided by the North-Rhine Westphalian and Bavarian forest laws: indeed the ordering of the latter's 1982 law in respect of the statement of forestry objectives follows the Federal Act. And, as Bauer (1987), notes, the prescriptions for other Länder are similar. Nevertheless the Länder differ in the degree of regulation they impose on private owners. Although many states do not require management plans, these being necessary only for purposes of income tax relief (see below), Hessen for example requires a plan for any private forest large enough to support a sustained yield (Richardson, 1985).

Another expression of the Federal government's role flows from its financial support of forestry. Every year, the Federal Minister decides with the Länder the measures to be promoted and the federal contribution (FAO, 1988; Giesen, 1991b), to which the Länder may add their supplements.

The most relevant point in the 1975 Federal Act concerning policy is Section 1 entitled 'Purpose of the law' which reads as follows:

> It is the object of the Act in particular:
> 1. to conserve forests for their economic value (economic function) and their environmental significance – the latter having a bearing especially on the long-term productive efficiency of the ecosystem, on the climate, on water resources, on the filtration of the atmosphere, on soil fertility, on scenic aspects, on agri- and infrastructure as well as on recreation for the general public (protective and recreation function); where required forests are to be expanded; their proper (*ordnungsgemässe*) management is to be ensured in a sustained manner;
> 2. to advance the forestry sector and
> 3. to strike a balance between the interests of society and the vested interests of forest owners.

Although the list of particular goals outlined in the all-embracing subsection **1** on environmental purposes is significant, the most important statement here is in subsection **3** since it applies whatever goals are identified. Next in importance is subsection **2** which must imply its own claim on government resources. It is arguable that subsection **1** is too general by being too comprehensive a statement in relation to the environmental effects of forests but the expansion and sustention themes are highly significant.

The following specific requirements are important to note:

1. Restocking is obligatory.
2. New planting requires the approval of up to four levels of administration including the agreement of agricultural and nature conservation authorities, and is generally not favoured in forest-rich areas.
3. Management is to be conducted in an orderly way (ordnungsgemäss is the German word with its own burden of meaning), and in a sustained manner.
4. Protective reserves can be designated for purposes of meeting conservation goals.
5. Access on foot to all forests is assured.
6. Various forms of cooperatives similar in style to the French 'groupements' are provided for.

A digression on the air pollution debate

The reference to 'orderly' or 'proper' management has taken on a special significance in the light of the debate on air pollution damage. There is a further reason for discussing the subject here. This is that in no other country has the war of words on air pollution damage, or supposed damage, to forests been more intense than in Germany. The debate has provided graphic proof of the concern of all Germans over the fate of their forests. To some, the evils forecast as a result of continued emissions of a whole cocktail of chemicals were such as to threaten the continued existence of forests throughout the country, indeed that was the firm belief of some very senior officials in the early 1980s. The discussion engaged the attention of the popular press and media generally as well as of politicians, highly interested foresters and forest scientists.

By 1992, however, a more balanced recognition of the extent and importance of the air pollution problem had emerged in contrast to the pessimism prevalent at the beginning of 1980s. A measure of the change in attitudes is provided by the publication of a report from the Federal Republic's Research Advisory Council in September 1986 (Forschungsbeirat Waldschäden (FBW), 1986) which said in its comment on Forest damage and air pollution:

> The FBW sees its statement of 1984 as still applying, namely that air pollution is a contributory factor in the new form of forest damage. An hypothesis which proves conclusively that damage occurs without the effect of emissions has not been produced. There is further proof that photo-oxidants (e.g. ozone) have a damaging effect and that the soil is being acidified by deposition from air pollutants and that this is indirectly affecting trees. These effects plus additional nitrogen from anthropogenic sources can destabilise ecosystems in certain stands. The FBW wishes to see air pollution reduced . . .

This laconic report was issued by the Federal Ministry of Agriculture with a covering statement noting that the research council 'confirms that forest damage is a matter of phenomena that are very complicated and hard to comprehend, on which air pollution has a substantial impact'. Later in the same year a conference called by the Federal German Society for Agricultural and Environmental Policy discussed two principal themes; one setaside, the other 'whether public concern about forest damage was hysteria or the expression of a genuine threat'. (An accessible statement of the scientific understanding reached by 1987 is provided by Rehfuess (1987).)

Finally, an unpublished report commissioned by the Federal Secretary of State for Research and Technology entitled 'Regional statistical analysis of the relationship between site conditions and forest damages' provide some astonishing conclusions. It must be borne in mind that (subjective) observations on crown thinness and leaf discoloration, the features which are used by too many scientists to signify 'damage', together with soil, climate and other data have been collected since the early 1980s under the auspices of the Economic Commission for Europe and that this set of forest stands had been supplemented by an EC network. There is, therefore, an enormous body of data on which to draw in order to study associations, or lack of them, between factors of the stand and site and damage observed. Three of the conclusions are particularly worthy of note:

1. For all species the soil has a great influence on damage. Significantly higher damage occurs on soils derived from magmatic rocks, nutrient-deficient substrates and rendzinas. Older soils which naturally tend towards acidity show less damage than rich soils. This result contradicts the hypothesis of soil acidification as a cause.
2. The dependence of damage on climatic factors is undoubted. It is probable that drought contributes to damage.
3. Data on SO_2 and deposition of NO_3 and SO_4 show no statistical connection with damage.

It is now clear that the scale of the actual, as opposed to presumed, damage was less than previously thought, that only a part could be ascribed to air pollution, and that the justification for the extreme predictions of some writers on the possible extension of damage and the extent of death of forests (for example, Nilsson *et al.*, 1991) was lacking. As Innes and Boswell (1990) pointed out, the role of air pollution is now seen as being extremely subtle in most areas. They noted that the widespread death of forests as seen in parts of eastern Europe (including, in this context, eastern Germany) as a result of high concentrations of sulphur dioxide combined with extreme winter stress had not occurred in western Europe. Furthermore, and confusingly, British experiments in open-top chambers indicate that for young trees there may be effects on growth (Lee *et al.*, 1990) without visible foliar damage.

One very positive result has emerged as a result of the fears generated in the course of the discussion on the so-called 'new form of forest damage'. Very important political agreements were obtained in the European Community on the control of emissions. These came about largely at the instigation of the German government which was heavily influenced by the depth of feeling on the possible effects of damage to the country's forests. The influence of foresters in the debate on damage underlying this series of political outcomes appears to be clear; whether the ethics of those engaged in the debate were always as professionally acceptable as desirable remains to be appraised.

The great environment/multiple-use debate

The Ministry of Agriculture issued an official statement on 'Proper Forestry' in part because it recognized that the term 'proper' referred to in the Federal Forest Act was not defined by law (Bundesministerium, undated). The statement sets out the characteristics of appropriate forestry. These are described, *inter alia*, as:

> securing sustained wood production and preserving forest ecosystems as a habitat for a great diversity of plants and animals by striving for healthy, stable and diverse forests; avoiding large-scale clear cuts; choosing species appropriate to the site while maintaining genetic diversity; developing the forest in line with requirements by conserving landscape, soil and stand to the greatest possible extent; undertaking operations with care, particularly in the case of regeneration, logging and transport; applying soil-preserving techniques, including the use of plant nutrients (as noted above); abstaining as far as possible from the use of pesticides and adopting as far as possible integrated plant protection methods; and, finally, aiming at game densities that are adapted to the forest and its regeneration.

The occasion of the statement was not, however, an intellectual concern with precision about the way in which forestry is practised. The main cause was the political strength of pressure groups wishingto see more forests designated as nature reserves from which all management would be excluded. This strongly expressed feeling has led to the growth of a minor industry discussing the aims of management and the form of silviculture to be adopted. The subject raised is a fundamental one and has a strong ideological basis. Its origins, though hardly its wilder manifestations, lie in the concern over whether the form of forestry developed and generally adopted in Germany and aimed at a productivist goal (cf. the discussion in Chapter 7 on France) is one which society as a whole favours in the closing decades of the 20th century.

Although most of the pressure has come from environmentalists, a substantial number of foresters and owners share the same ambitions. Indeed a forestry group, the Arbeitsgemeinschaft Naturgemässe Waldwirtschaft (ANW), including some enthusiastic private owners, which has existed for over 40 years, promotes the ideas of near-to-nature forestry. It is also the seed from which the organization Pro-Silva has sprung. The strong liking for order in all walks of German life is matched in forestry by a systematic type of management and a defensive attitude to any change proposed from outside the forestry sector. The demand for change toward a greater concentration on the non-material products of the forest is thus a testing one for the profession.

To some extent the argument covers old ground since part of the reasoning centres on the sustainability of what have come to be regarded as conventional forestry practices. An important benchmark was the study undertaken by Wiedemann in Lower Saxony on the soil impacts and subsequent damage to the health and vigour of Norway spruce plantations in the early 1920s (Troup, 1952). The profession as a whole is engaged with the debate: only a brief view of the forestry arguments is possible in a field which contains a rich literature. The prevailing attitude of many foresters is well expressed in, among others, Bauer (1987), Brandl (1991), Köpf (1990), Sprich and Waldenspuhl (1991). A composite of the various foresters' contributions provides an argument as follows:

1. Forest management is necessary to provide the environment for man's enjoyment of the forest and for the protective functions the forest performs (this is the well-known 'wake' theory).
2. Such management is in fact close to nature, since native species are or should be used and environmentally friendly silvicultural systems and operational techniques are used.
3. The cost of such management can be supported by growing trees for the market.
4. Therefore active management for wood production must continue.
5. However, the chances of making this activity pay would be greatly reduced were foresters to respond to the clamours of those who wish to see more attention paid to nature conservation leading at the extreme to a ban on management.

Rehearsals of the arguments were frequently heard in 1990 and 1991 after the Federal Constitutional Court made a remarkable statement in May 1990 about the role of public forests in Germany (Sturm and Waldenspuhl, 1990; Huber-Stentrup, 1991). The Court's remarks appeared in a tailpiece to a judgement concerning the tax liability of a particular fund. Part of the reason for the remark that, freely translated, 'communal and state forests serve environmental and recreational needs and not the assurance of sale and use of forest products' was to provide a basis for the judgement which set forestry apart from agriculture. Considered from outside Germany, perhaps the most remarkable aspect of the statement was the extent of the failure of communication between those responsible for forestry and the legal authorities working in the highest court in the land.

Despite the fact that the need for an objective method of balancing the values of the various outputs is well recognized (see Volz, 1991), remarkably few objective appraisals such as cost–benefit assessments have been attempted (Bergen, 1991 refers to some of the studies made in forestry faculties on this subject). Grammel (1987) points out that rational hierarchies of objectives may be replaced by a system of ideologies. More positively, Brabänder (1987, 1990) has proposed that the restrictive regulatory framework controlling owners' actions must be dismantled before it will be possible to achieve an active response to the demands of those who wish to enjoy greater recreation and protection outputs from forestry.

The nature conservation issue interacts with hunting: the creation of the untouched mixed age and species forest desired by environmentalists does not provide particularly good deer habitat because of the lack of open regeneration areas. As pointed out by the Bundesminister (1991a), the cull of deer is no longer determined by the stock of animals but by the damage wrought to regeneration and to standing trees. Most tree growers would argue that the stock level remains far too high; they in turn are obliged to adopt high cost fencing and individual tree protection.

Administration

Within government the relevant authorities are the Federal Ministry for Food, Agriculture and Forestry and, at 'Land' level, the respective Ministries of Agriculture or of the Environment. The former is concerned mainly with general policy, relations with other departments and external relations. In some northern Länder the function of implementing policy towards private forests is undertaken by Chambers of Agriculture, elsewhere by the Land forest service itself. Forest services are responsible for the administration of hunting.

There is a large number of societies and councils concerned with forestry in Germany including scientific, professional and commercial bodies. So far as private forestry is concerned the most important is the Association of German Forest Owners (Arbeitsgemeinschaft Deutscher Waldbesitzerverband) which represents the interests of both private and communal forest owners' associations in the individual Länder. The German Forestry Council (Deutscher Forstwirtschaftsrat) has a special place in the development of policy and its administration since it provides a focus for discussion by representatives of all ownerships together with professional and scientific wings. The Council is normally concerned with issues of topical importance, the conclusions of its deliberations being usually addressed to government for action.

Implementation

Cooperatives

The 1975 Federal Forest Act integrated the provisions covering a number of kinds of cooperatives. This development occurred as a result of general political agreement in favour of the maintenance of a broadly based distribution of forest properties and the desire to promote cooperation among owners (Oedekoven, 1981). Voluntary cooperatives play an important economic role among the smaller ownerships, and strong claims have been made about the extent of gains in income as a result of their activity (e.g. Brabänder, 1981).

TAX

The most important tax provision concerning forestry is without doubt the assessment of capital value through the valuation known as 'einheitswert' (termed here standard value) (Kroth, 1976). This was introduced in its present form in 1934 for the assessment of agriculture and forestry. The principle in relation to agriculture is that for every soil type and climatic region it is possible to predict the appropriate crop and level of yield to be expected under good management (Weiers and Reid, 1974). From this it is easy to calculate the net income which, when capitalized, establishes the capital value. However, subsequent appraisal of the actual yields of agricultural crops has shown that the correlation with the land types distinguished when the tax was set up was very low. In forestry the route is via the chosen species and its predicted yield class in the particular 'gemeinde' or district (Anon., 1967; Bachofer, undated; Kroth, 1963). Three capital taxes, namely land, inheritance and wealth, are based on this assessment of capital value (Schödel, 1981). These values were last reviewed in 1964 when the 1934 rate was left unaltered. Quite apart from broader fiscal policy considerations, such as the political acceptability of increased tax burdens on agriculture and forestry, the general unwieldiness of the system, the exceptions which have to be made (see below) and the dubiety of the exactness of the relationship between yield and land quality noted above must all call into question the value of this basis of a property tax.

Income tax

This is the most burdensome of direct taxes bearing on private forestry. Marginal rates of tax vary (1991 figures) between 19% and 53%. For a private woodland owner holding the average size of property of some 10 ha (cf. Table 10.1, properties over 1 ha) and allowing for the fact that he has other sources of income, the average rate of income tax is taken to be 30%. With an average net cash income of the order of 100 DM ha^{-1} (Anon., 1991a, Table 47) the charge becomes $0.3 \times 100 = 30$ DM ha^{-1} per year. This would be reduced if the cut in a given year were above the normal yield owing to storm or other damage thus requiring a higher rate of removal of timber (as explained in Korsch, 1990). Accounts are required for properties over 30 ha where the owner wishes to claim the various reliefs from income tax arising with extraordinary revenues and revenues from windblow.

Land tax

Land tax is calculated on the basis of the forest's standard value. The calculation requires multiplication by a tax rate, currently 0.6%, and lastly by a rate which is currently 265% for forestry businesses. Thus for a typical property for west Germany as a whole with a 'standard value' of 200 DM ha^{-1}, the charge amounts to $0.006 \times 2.65 \times 200 = 3.2$ DM ha^{-1} per year.

Wealth tax

This tax is levied at a rate of 0.5% per annum on net wealth of individuals above 70,000 DM. Assuming other assets of moderate value beside the forest, the average tax rate would be of the order of 0.4% and the liability in respect of the forest would thus be 0.004×200 or 0.8 DM ha^{-1} per year.

Inheritance tax

The tax is highly progressive with 1990 rates of from 3% to a maximum of 70%. There are various methods open to testators to reduce the burden of tax on beneficiaries. It is assumed here that the average rate applying to a transfer to a child is 10%. With an average interval between deaths of 30 years, and standard value of 200 DM ha^{-1}, the annual equivalent charge is $0.10 \times 0.0246 \times 200$ or 0.5 DM ha^{-1}.

Summary of direct taxation charges

Total tax payments estimated on a countrywide average basis are as follows (DM ha^{-1} $year^{-1}$):

- Income tax 30
- Land tax 3.2
- Wealth tax 0.8
- Inheritance tax 0.5
- Total 35

GRANTS

With exception of the special case of afforestation of setaside land, grants provide only a relatively small element of aid to forestry in West Germany. In 1989, including support of this work, they amounted to nearly 120 million DM, 60% coming from the Federal government. For the whole of the Federal Republic in 1989 (Thoroe, 1991) the distribution across operations benefitting was: silviculture 29%, fertilizing as a remedial measure against acidification and remedial restocking 56%, roads 13%; and formation of cooperatives 2%. No grants are payable in respect of normal restocking. In addition to these grants, Länder provided a further 31m DM out of their own resources in 1988 (Schwoerer-Böhning, personal communication). Grants are highly differentiated both by activity and region within a Land, so that in 1991 the scheme for Bavaria included 36 separate rates across 12 operations.

Setaside

Planting grants vary between and within Länder. In most states, rates are set in absolute amounts and differentiate between broadleaved, mixed broadleaved and conifer and purely conifer planting (Plochmann and Thoroe, 1991). Separate grants are paid for beating up and weeding. The rate for planting averaged 6000 DM ha^{-1} across western Germany in 1990 (Gallus, 1991). From 1991, the farmer was to receive a premium payable for 20 years of 500 DM ha^{-1} per year for planting of setaside land (Bundesministerium, 1991b).

Extension

Extension is the responsibility of Land governments, and the authority, either forest service or chamber of agriculture, is empowered to provide extension services free. No central data are compiled on the costs of extension work or the man-years of professional and subprofessional personnel engaged. The indication from data in two Länder on forester numbers and the amount of time devoted to advice and administration of grants is that the advisory role absorbs a very large volume of resources, perhaps approaching the spend on grants themselves.

Research

The EC survey return for research shows the total man-years of staff personnel engaged in West Germany as 348. Assuming an average expenditure of 120,000 DM per worker-year and making a 10% addition for forest products research implies total expenditure of some 46 m DM. Consideration of the allocation of such expenditure to different ownership categories, showed that private forests accounted for 39.4% of the forest area of the FRG and 35.2% of the cut over the years 1987–1989. Assuming that public forests enjoy a somewhat higher share of the benefits of research than the dispersed private estates and the many farm forests, a share of 30% of the total cost of research appears a plausible allocation of cost, implying expenditure of 14m DM.

Total Public Expenditure in Support of Private Forestry (FRG)

Total expenditure estimated for 1990 was as follows (million DM):

- Grants, including afforestation 151
- Research 14
- Total, excluding extension 165

Commentary on Implementation Measures

Apart from the provisions for income tax, the charge for which is substantially moderated where cutting is determined by circumstances beyond the owner's control, taxation is generous towards forestry. The control exerted by supervision, grant administration and the like is marked. Against this, the effort devoted to extension in the broadest sense is perhaps as intense as in any other major forestry country.

Outstanding Issues

Farm forests are recognized as a distinct category in relation to the policy towards private forestry, taxation and statistics. The problems of the small-scale seen in Bavaria, for instance, will no doubt remain with any change in the Common Agricultural Policy. The point was made in the German statement at the 1991 meeting of the FAO European Forestry Commission that afforestation of farmland provided an opportunity to improve the distribution of forests over the country. While true, this fact offers no help to owners of farm forests most of whom work in already well-forested regions where increased forest cover is opposed by the public and specifically played down by the 1975 Forestry Act (Section 6, sub-section (3), Federal Ministry, 1984).

It is clear that the discussion on the form of forestry to be adopted will continue for as long as there are zealots on both sides to argue about it and there is a market for wood, since this provides such an important part of the finance to maintain traditional forestry operations. In practice, adjustments will increasingly be made as a result of political decisions, but it is clear, as Otto (1991) points out, that there is a need for a meeting of minds of foresters and environmentalists. Extension services have a major responsibility for education of the public. Decisions can be expected to influence the designation and management of public sector forests first, but there will be certain parts of private forests which because of their location, biological interest or other special factors will rank for new types of treatment. A more rational basis for such decisions is needed and economic and social research should, despite the negative views of some (see, for example, Köpf, 1990), assist in this process. There is certainly a strong incentive for private owners to press for more objective assessment.

The rules for appropriate forestry incorporate an injunction to safeguard the forest from animal damage as they do for protection against the supposed effects of air pollution. There is an amazing toleration of game damage compared with that ascribed to the other agency. Forest services have an unenviable role as they are also the hunting authorities. The hunting lobby is strong and many private owners form part of it or benefit from it in terms of rents; by contrast everyone believes that his woods suffer or are liable to suffer from air pollution damage. In effect hunters use the forest as a low cost habitat for their sport. Even if a thorough monetary evaluation of hunting were made which showed current activity to be excessive in terms of joint benefit from wood production and hunting, it is unlikely on present form that this would lead to a diminution of stock and damage, so great is hunting's attraction.

References

Anon. (1967) [Regulation for application of Section 55 sub-sections 3 and 4 of the valuation law (forest valuation).] *Bundesgesetzblatt* no. 46, 805–836.

Anon. (1991a) [Table 94 from section III. Forestry and timber industry.] *Deutscher Bundestag* 12th Session, Bonn.

Anon. (1991b) [Forestry and the timber industry.] *Deutscher Bundestag* 12th Session, 63–70 *et seq.* Bonn.

Anon. (1991c) [*Management Results for Different Ownership Categories in Baden-Württemberg, Forest Year 1990.*] Forest Research Institute, Freiburg.

Bachofer, W. (undated) [Analysis of standard value statistics 1964 – forestry.] Typescript.

Bauer, O. (1987) [State forest management between economy and ecology.] *Allgemeine Forst Zeitschrift* 42(1/2), 3–5.

Behrndt, M. (1989) [*Influence of Wood Markets on the Yield Position of German Forestry*]. Doctoral dissertation, Göttingen.

Bergen, V. (1991) [The forest as a factor of the economy]. *Allgemeine Forst Zeitschrift* 46(1), 17–19.

Brabänder, H.D. (1981) Subsidies and efficiency in forestry cooperatives. In Järveläinen, V.-P. (ed.), Effectiveness of forest policy on small woodlands, *Silva Fennica* 15(1), 79–84.

Brabänder, H.D. (1987) [Crisis or change in forestry?]. *Der Forst-und Holzwirt* 42(14), 367–370.

Brabänder, H.D. (1990) [Possibilities and limits of use of money incentives for securing landscape requirements in the forest]. *Forstarchiv* 61, 242–247.

Brandl, H. (1987) [The history of the business side of forestry]. *Allgemeine Forst Zeitschrift* 40/41, 1019–1023.

Brandl, H. (1991) [Economic possibilities for forestry.] *Der Waldwirt* 18(8/9), 131–139.

Bundesministerium für Ernährung, Landwirtschaft und Forsten. (Undated) Annex 4. *Proper Forestry*. Bonn.
Bundesminister (1991a) [*Our Forests. Forestry and Timber Industry in Germany.*] Bonn.
Bundesminister (1991b) Forestry grants. *Agrarpolitische Mitteilungen* no. 10/91. Bonn.
Crowther, G. (1940) *An Outline of Money*. Nelson, London.
ECE/FAO (1992) The forest resources of the temperate zones. The UN-ECE/FAO Forest Resource Assessment, *Volume 1: General Forest Resource Information*. UN, New York.
Eisenmann, H. (1983) [Basic principles and aims of Bavarian support for agriculture and forestry.] *Förderungsdienst* 31(12), 365–370.
FAO (1988) Forestry policies in Europe. *FAO Forestry Paper* 86. Rome.
Federal Ministry of Food, Agriculture and Forestry (1984) Federal Forest Act of 1975 (trs. W. Ellers). *Bundesgesetzblatt*, 1037.
Forschungsbeirat Waldschäden (1986) 2. [*Report*.] Karlsruhe.
Fulbrook, M. (1990) *A Concise History of Germany*. Cambridge University Press, Cambridge.
Gallus, G. (1991) [*Forests and Forest Policy in Germany*.] Press release, Der Bundesminister für Ernährung, Landwirtschaft und Forsten, 22 August. Bonn.
Giesen, K. (1991a) [*Farm Forestry*.] Circular 4/91. Arbeitsgemeinschaft Deutscher Waldbesitzerverbande (ADW), Bonn.
Giesen, K. (1991) [*Forestry grant aid 1991*.] Circular 11/91 covering extract from German Parliamentary report-12th session: Ground rules for forestry grants. ADW, Bonn.
Grammel, R.H. (1987) Forestry utilisation: ideology and practice. *Forstwissenschaft Centralblatt* 106(3), 182–186.
Greeley, W.B. (1953) *Forest Policy*. McGraw Hill.
Huber-Stentrup, E. (1991) [The misunderstanding Constitutional Court.] *Allgemeine Forst Zeitschrift* 46(8), 413–414.
Innes, J.D. and Boswell, R.C. (1990) Monitoring of forest condition in Great Britain, 1989. *Forestry Commission Bulletin* 94. HMSO, London.
Klose, F. (1985) *A Brief History of the German Forest – Achievements and Mistakes Down the Ages*. GTZ, Eschborn.
Köpf, E.U. (1990) [Forest policy today.] *Allgemeine Forst-und Jagd Zeitung* 161(6/7), 113–115.
Korsch, J. (1990) [Ground rules of forestry taxation.] Forstliche Versuchs-und Forschungsanstalt Baden Württemberg. Typescript. Freiburg.
Kroth, W. (1963) *Systems of Forest Taxation and Tax Impact upon Private Forestry in Several European Countries*. (trs.Israel Program for Scientific translations.) Jerusalem.
Kroth, W. (1976) Systems of forest taxation and the tax liability of private forest holdings. In: *Forestry Problems and their Implications for the Environment in the Member States of the E.C.* Information on Agriculture No. 25. Brussels and Luxembourg.
Lee, H.S.J., Wilson, A., Banham, S.E., Durrant, D.W.H., Houston, T. and Waddell, D.A. (1990) The effect of air quality on growth. *Research Information Note* 182. Forestry Commission, Alice Holt Lodge, Farnham.

Niesslein, E. (1991) [The Federal Constitutional Court is wrong!] *Allgemeine Forst Zeitschrift* 46(8), 408–412.

Nilsson, S., Sallnäs, O., Duinker, P. (1991) *Forest Potentials and Policy: a Summary of a Study of Eastern and Western European Forests by the International Institute for Applied Systems Analysis.* Executive Report 17. Laxenburg.

Oedekoven, K. (1981) Small private forests in the Federal Republic of Germany. *Journal of Forestry* 79(3), 161–162.

Ott, W. (1987) [The money yield of forestry]. *Allgemeine Forst Zeitschrift* 16/17, 399–401.

Otto, H.-J. (1991) [Forest ecology, silviculture and nature conservation.] *Allgemeine Forst Zeitschrift* 46(1), 9–14.

Plochmann, R. and Thoroe, C. (1991) [Support of afforestation: cost-benefit analysis of afforestation support.] *Schriftenreihe des Bundesministers fur Ernährung, Landwirtschaft und Forsten, Reihe A*: Angewandte Wissenschaft, Heft 397. Landwirtschaftsverlag, Munster-Hiltrup.

Rehfuess, K.E. (1987) Perceptions of forest diseases in central Europe. *Forestry* 60(1), 1–11.

Richardson, D. (1985) *The Role of the State in the Regulation of Yield from Private Forests.* New Zealand Forestry Council. Working Paper No. 4, Wellington.

Ripken, H. (1991) *Planning, Control and Management System.* Paper to Joint FAO/ECE/ILO Committee workshop on the organization and management of forestry under market conditions. Budapest 30 September–4 October.

Rozsnyay, Z. and Schulte, U. (1978) [The draft of the National Forest Law (1942) and its significance for later West German forestry legislation.] *Schriften aus der Forstlichen Fakultat der Universitat Göttingen* no. 60.

Schödel, H. (1981) [The forest and taxes.] Address to tax seminar, Agriculture Office, Munich. Typescript.

Sprich, L. and Waldenspuhl, T. (1991) [Ecology and economics.] *Allgemeine Forst Zeitschrift* 46(1), 15–16.

Sturm, K. and Waldenspuhl, T. (1990) [Basis of the judgement of the Constitutional Court on management goals in public sector forests.] *Allgemeine Forst Zeitschrift* 45, 1146–1148.

Thomson, D. (1966) *Europe Since Napoleon* 2nd edn, Penguin Books. London.

Thoroe, C. (1991) *Benefit-Cost-Analysis for Afforestation of Agricultural Land.* Paper to EC Workshop on afforestation of agricultural land, Brussels, December 1991.

Troup, R.S. (1938) *Forestry and State Control.* Clarendon Press, Oxford.

Troup, R.S. (1952) In: Jones, E.W. (ed.), *Silvicultural Systems*, 2nd edn Clarendon Press, Oxford.

Volz, K.-R. (1991) [Near-to-nature forest management in stormy times – a forest policy problem formulation?] *Holz-Zentralblatt* 16 August, 1508–1509, 21 August, 1525–1526.

Weiers, C.J. and Reid, I.G. (1974) Land valuation and taxation: the German experience. *Centre for European Agricultural Studies: Miscellaneous Studies* no. 1. Wye.

Zon, R. and Sparhawk, W.N. (1923) *Forest Resources of the World*, Vol. 1. McGraw-Hill, New York.

11

SWITZERLAND

Background

Six facts stand out as being of fundamental importance for any discussion of private forestry in Switzerland.

1. Because of the pervasive risk of damage to settlements and land by torrents and avalanches, careful maintenance of soil cover and man-made structures is necessary in order to protect soil and to control watercourses.
2. Managed forests are held to play an important role in achieving the requisite measures.
3. Private forests amount to 316,000 ha, or 26% of the total forest area, and are in the hands of more than 250,000 owners.
4. Since 1982 public forests in the Alpine region, and in more recent years elsewhere, have been run at a loss (Peter, 1989) and standing values of spruce in real terms have fallen more than 75% over the last 25 years (see below).
5. The annual cut has averaged 4m m^3 over the past 30 years, and may be compared with a level of desirable cut of 6.5m m^3 or more (OFEFP, 1989): the growing stock now averages 333 m^3 ha^{-1}.
6. Federal support of federal, cantonal and private forests by way of grants and investment credits rose from 127m Swiss francs (Sw.fr.) in 1988 (FAO, 1988, p. 232) to 270m Sw.fr., or a remarkable 250 Sw.fr. ha^{-1} of all forest, in 1991.

The dominant concern of Swiss forest policy and its careful practical application is that of protection of the country's land surface from loss or damage through erosion, avalanches and torrents. Minimization of damage from these influences requires careful attention to the design of all man-made structures, the arrangement of land drainage, the conduct of agriculture and the form of silviculture that can be applied or it is thought necessary to apply. There are strong controls over deforestation and over the location of compensatory afforestation where permitted works require woodland to be cleared. Also, because there is a stringently applied requirement not to expose the soil, especially in mountainous areas, selection forest of one form or another occupies 17.3% of the total forest area.

There is a further point which is crucial to any assessment of the current position of forestry in Switzerland. This revolves around the question of just what constitutes the best silvicultural recipe to provide the desired cover and protective effects. Official Swiss publications (for example OFEFP, 1989) raise the question 'Are the forests too old?' Swiss forests contain huge levels of growing stock: the average is easily the highest for any European country. In part this stems from the species composition and the generally favourable conditions for tree growth. The average age of the long-managed (not old growth) high forest is 74 years and increasing. There are two concerns. One is that the age-class distribution shows a high proportion of the older classes and too small a proportion of the young actively growing stands considered necessary if the forest's protective function is to be maintained. The situation is growingly regarded as alarming, with the prospect of increasing wind and snow damage among over-age stands, greater difficulties of logging broken or otherwise damaged material, clogging of watercourses, etc. The second is that the position is likely to worsen since the cut is only 60% (Price, 1988, quoting Rechsteiner) of the increment and some 61% of the sustainable yield (OFEFP, 1989). Increment as a proportion of the growing stock at 1.6% is now as low as that of any other European country outside the Mediterranean region (ECE/FAO, 1992).

Clearly, the question of how near the age distribution is to the optimum cannot be regarded as solely a technical one: in fact it is principally an economic one since the usual signals that have governed management have in effect been reversed during the 1980s. The main problem is the declining profitability not only of forestry taken as a whole but of logging itself. It is little to be wondered that the government statement (FAO, 1988, p. 231) should have emphasized the matter thus:

> The economic situation of forestry is becoming increasingly difficult and there are more and more forest owners who find that the income from the forest no longer covers expenditure. The consequences vary according to the circumstances of the owners. There is a risk that the level of management will suffer over very large areas, especially in the mountains.

The Financial Position of Forestry

Several sets of data are valuable in reviewing the economic position of forestry. One measure is seen in the figure for declining net revenue (Office

Fédérale de la Statistique, 1991). The data shown in Table 11.1 refer only to public, i.e. communal and federal, forests: these account for 73% of the total forest area. (Although certain economic results are estimated for private forests, figures are not compiled for net cash outturn; a sample of private forests was drawn in 1991 in preparation for collection of the requisite data in future.) For wood-producing forests in the public sector as a whole net annual income flow fell almost to zero in 1988 and 1989, while for the three years 1987–1989, net receipts from the total area of public forests have been negative.

The values given in Table 11.1 cover silvicultural and protection operations as well as the harvesting of wood which is by far the largest source of revenue in the forests covered in the tabulation. Peter (1989), sets out the time trend of net income by region, showing how both the Jura and Mittelland are approaching the difficult position of the Alps, and provides detailed evidence on logging costs and prices of assortments for 1988. It is the trends in the costs and returns from harvesting that are most relevant. Table 18 of the Statistical Office report on forestry allows the standing value of trees to be derived, direct assessment of standing prices being impracticable because the great bulk of wood worked in Switzerland is sold felled. For the period 1971–1989, it appears that average standing values in real terms fell by almost 60% in the Alps but by only one-third in the Plateau region.

Such data refer to average results and the composition of the cut may well have differed in species, size, location, etc. over the years. It is therefore useful to note some more standardized information compiled by BUWAL (1991) which relates to log prices and costs of harvesting. Table 11.2 shows costs and prices for Norway spruce.

The results in Table 11.2 suggest, as would be expected, that for a given tree size the position has been even worse than that for the aggregate of all harvesting. There are two principal causes of the decline in cash returns from forests. The first, and by far the most important, is the very rapid decline in prices of logs at roadside over the past three decades and especially since the mid-1970s. When deflated, the values shown for the price of spruce wood at roadside indicate a fall in real terms of 57% from 1964 to 1989. The second is that rates of increase in labour productivity in both harvesting and forestry

Table 11.1. Overall cash outturn for public forests, selected years.

	Millions of Sw. fr.			
	1975	1980	1985	1989
For all public forests	44	113	12	−14
For enterprises producing wood	99	168	64	2

Source: Office Fédéral de la Statistique (1991), Table 15.

Table 11.2. Calculated standing values for spruce.

	Prices and costs (Sw.fr. m^{-3})						
	1964	65	70	75	80	85	89
1. Long logs, 3rd. cl., roadside	107.3	105.8	108.9	135.2	140.1	122.8	119.2
2. Pulpwood, roadside	54.7	54.7	56.0	92.0	91.2	84.8	79.2
3. Weighted roadside price	97.3	96.1	98.9	127.0	130.8	115.6	111.6
4. Harvesting cost	19.7	20.2	23.8	34.5	42.2	57.5	64.9
5. Derived standing value	77.5	76.0	65.0	92.5	88.7	88.2	46.6
6. Deflated value	69.9	64.9	47.7	45.4	36.9	20.1	15.8

Source: author's calculation, based on: rows 1 and 2—prices from return for 1991 ECE/FAO study of roundwood prices and specifications; row 3—weighted average price based on weights for conifer cut in the 1970s and 1980s, namely 80% sawlogs, 19% pulpwood; row 4—from official return (the series for this measure is unavoidably less certain and in fact represents broad average values for all species); row 6—calculated using the consumer price index as deflator.

generally have not matched the increase in level of wages, which is as high as anywhere else in the world. (The World Bank, 1990, estimates Switzerland's GNP per caput as the world's highest at $30,270). The data for logging cost in Table 11.2, when deflated, indicate a rise in real cost of some 23% between 1964 and 1989.

Parts of Germany show a similar trend in net annual income from forests which are in any sense going concerns, but the unwelcome developments revealed by these reports for Switzerland, and especially those for the Alpine region, are not parallelled in a market economy anywhere else in the world and certainly not in any country with long, consistently applied management. This untoward position in a forest estate which is well roaded and managed calls for decisive action if it is to be solved commercially rather than by subsidization at an increasing rate. Whether the newly introduced special programme of training for workers will prove more than a palliative is doubtful.

Although the financial position is not known for private woodlands there is little reason to think that it is not as difficult, or will not shortly become as difficult, as in public forests. With some 250,000 owners of private woods (Office Fédéral de la Statistique, 1991) the average size of ownership at about 1 ha is very small; over 90% of owners hold less than 10 ha (Swiss Federal Office for Forestry, undated). Up to recent times such woods have remained predominantly in the hands of farmers. Increasingly, however, private woods are owned by people other than the farmer of the surrounding land; 58,000 farmers accounting in 1989 for only 40% of the private forest area. Not only does this complicate management, but it also means that the cheap harvesting once practicable by the farmer using his own tractor and labour at a slack time in the farming calendar is no longer possible (Amstutz, personal communication). In addition, labour is no longer readily available, especially in the more remote areas. Despite this position and the powers of the 1902 Forest Police Law in relation to the discouragement of fragmentation and the encouragement of consolidation of parcels of woodland, these have had no effect. It is noteworthy that the situation does not excite the same concern as observed in France over the past 40 or more years.

Current Policy and Legislation

The government summary of forestry policy objectives at the Federal level set out in FAO (1988, page 232) is bland in the extreme, thus:

1. The conservation of the total area under forest and its general distribution.
2. The development of the multiple functions of the forest.

3. The promotion of measures for the protection of the countryside against natural disasters.

Although the same source notes that forestry in Switzerland is extremely decentralized, a feature to be expected in the light of the traditions of fierce independence of the cantons, the same source makes it clear that federal legislation in forestry provides the framework for the more detailed laws and decrees enacted by each canton. Cantonal legislation may make goals more precise but the principal concern at this level of government lies with implementation.

The roots of current Swiss forest law can be traced in some places at least to medieval times. Their purpose was originally the protection of the forest as a vital instrument in soil and water conservation including flood control. As noted by Siegfried (1950) the Alemanni gave only a limited holding to each family and the rest of the land was held in common. Forests were so held and an administration on democratic lines was the natural result of this association of free men, this in turn being the basis of the commune. A typical 'Bannbrief' (Price, 1988 discusses these instruments for regulating the use of the woods of a commune), though one that is quite exceptional in terms of its date, is that from Andermatt dated 1397 (Swiss Federal Office for Forestry, 1987) which states:

> To all those who shall see or hear this proclamation being read – We the Councilmen 'on the Matt' in the valley of Urseren do hereby declare that we have decided after due thought and consideration to preserve the forest above the Matt and the plants within this forest for our own and future generations. Thus no one shall cut down or carry away either branches or twigs or plants or shrubs or cones or any other thing that may or has grown in this forest. Whosoever shall do so be he man or woman old or young or whosoever shall hear of such a deed being done at any place or any how that person whoever he may be shall bear a debt to each and every Councilman of five pounds in pennies of whichever currency is in common usage at that time.

From 1530 onwards such Bannbriefe were enforced by specially appointed wardens (Price, 1988, quoting Tromp). Wide-scale awareness of the importance of forest for protection as well as for wood only grew towards the end of the 18th century. Thus, as requirements for food and wood increased, so the pressure on the forests rose and although positive measures were instituted in particular cantons, the position in terms of forest area, protection from flood and avalanche and availability of fuelwood had become so serious by the early 19th century that cantons took control of the legislation and a number of laws were passed to restrict felling, and to protect and improve forests.

Price (1988), in a succinct review of legislative developments over seven centuries, draws attention to the manner in which the federal concern for appropriate legislation and means of implementation has grown over the past 150 years.[1] The immediate causes of worry in the 19th century were severe floods; they were instrumental in the passage of the crucial forest law of 1876 which introduced the classification of forests into protective and non-protective, the embargo on felling in protective forests without federal authority, the injunction not to diminish the area of forest, effectively on a cantonal basis, and implying restocking or compensatory afforestation nearby, the power to regulate grazing and to require the afforestation of land in order to establish protective forest and the prohibition of sale of public forests. The 1902 Forest Police Law, like its predecessor, was so named because it regulated forest use. The two laws did not require a minimum level of forest management, a need that is now pressing because of the ageing state of the growing stock and its reduced capability for resisting storm damage and supplying beneficial forest influences.[2] Although some cantonal laws require that management be applied so as to increase wood yields, this is rare. Management plans, required by law in public forests, are not obligatory in private ones.

Today's definition of 'forest' in Switzerland is minimalist; an area of 0.05 ha with a minimum breadth of 12 m is included. This is relevant in relation to the satisfaction of conservation goals, since forests, which are typically of native species, are of higher conservation interest than most forms of farm land. It is hardly surprising that another matter relevant to policy discussions in the country is the desirable balance of wildlife conservation and wood production and, associated with this, the competence, in the legal sense, of the forestry authorities in decisions concerning the location of forests and their management. These have come to a head in the debate over a new forest law to replace the 1902 and several later laws.

PROPOSED LEGISLATION AND THE REFLECTION OF POLICY ISSUES

The concerns of policy in the early 1990s are threefold; first, the protection of the forest itself from damaging agencies, notably the new phenomenon of air pollution; second, recognizing the serious financial position of forestry, the need to find ways and to provide the means to ensure the

[1] Federal laws from the 1902 Forest Police Act onward are set out in the document 'Forêts' of the Chancellerie fédérale (1990).

[2] Certain of these long-established principles stand in contrast to those favoured by the protagonists of the 'New Forestry' in the north western States of the USA referred to in Chapter 16.

minimum level of management required to maintain the forest's protective role; and third, the desire to develop the other functions of the forest, namely recreation and wood. Several measures have been taken to reduce harmful emissions; these are described in FAO (1988, pp. 235–236). Provisions for achieving the more purely forestry aims are being addressed in a major revision of the law (see FAO, 1988, Conseil fédéral suisse, 1988; Conseil des Etats, 1988; and 1991; Price, 1988). The first matter covered concerns deforestation and compensatory works. The second relates to means of encouraging adequate management to achieve and maintain the forest's protective role.

From the point of view of policy aims, the new Bill is significant on account of three features. First, it proposes that private forest be regarded in every material, managerial way as equivalent to public forest, that is with the same obligations. (By contrast, the situation in Britain whereby private owners are obliged to restock after felling while until 1992 there was no such obligation on the State forest enterprise appears contrary.) The only interpretation of the strict rules of policy enshrined in the original Forest Police Act of 1902 and in all subsequent legislation is that a rigorous, even rigid, policy is essential if discipline over an important aspect of environmental policy is to be established and maintained. This appears unexceptionable.

Second, it is proposed that in future all Swiss forests should be regarded as protection forests.

Third, while retaining the long-standing requirement that the forest area of Switzerland should not be diminished, the Bill introduces some novel considerations. One is that forests should become subject to the legislation controlling land use. The second is that, in the case of deforestation, compensatory planting may be considered in a different region. (This subject is discussed below under 'Outstanding Issues'.) A third proposal introducing the principle of money compensation rather than compensatory planting in kind was dropped between the original and 1991 versions of the Bill.

Policy on forest management

The proposed law introduces proposals under the heading of management, including:

1. Working plans to be prepared and confirmed by the cantonal authorities.
2. Cantons to undertake necessary operations where a private owner fails to carry out prescriptions.
3. Most importantly, financial support up to 80% of the expenses of silvicultural, felling or extraction operations the costs of which are not otherwise covered.

Training

Recognizing the difficult situation in logging as well as other operations arising from uncompetitive costs, the government has initiated a new programme for training of forest workers and supervisors: this is termed PRO(fessionnelle)FOR(mation) (Buchel, 1991).

Administration

Officially, forestry in Switzerland is extremely decentralized and, as recorded in FAO (1988), the Federal government has only very limited rights in forestry matters. Nevertheless the federal authority, the Federal Forest Agency (Direction fédérale des forêts) has considerable influence through its function of financing forestry in the cantons. The Agency forms part of the federal department of the environment, forests and countryside. noteworthy feature of the Swiss forestry scene is the close linkage between federal, cantonal, communal authorities, their forestry officials and private owners and forest managers. This happy partnership may, however, be increasingly tested over the next few years. As in Britain, where central government support of local authority spending rose from three-fifths before the introduction of the poll tax to 80% in its aftermath, so in Switzerland federal support of forestry at cantonal and municipal levels of government increases. Prats Llaurado (1991) observed that among developed countries, Switzerland is the exception in relation to the general shift towards greater delegation of power and wider democratic participation. In the Swiss case, the rises in costs of forest management have led to increased central financing and control.

The Forest Agency is responsible for hunting. Private woodland owners have no hunting rights *qua* owners, these being allocated by cantonal and communal authorities for a particular number of head.

A single body, the Swiss Forestry Association (Waldwirtschaft Verband Schweiz, or Economie Forestière Association Suisse), brings the various levels of government forestry administration together with private owners in one organization. Its priorities are to develop policy, to promote the marketing of timber, to improve operational efficiency, and to provide members and the public with information. It produces a number of journals targeted at various membership groups (professional foresters, manual workers) and establishes recommended log prices for the succeeding felling season.

Implementation

Apart from the regulatory powers of the state, usually operated through the cantons, extension services, grants, investment credits and tax relief are available. Despite the growing financial difficulties of owners, federal funding has not expanded sufficiently to cover the compensation required to meet new obligations, such as the uneconomic harvesting favoured in the proposed law.

Tax

There is a multiplicity of taxes on forestry in Switzerland, most of which differ among the 26 cantons. In most cantons and municipalities, income and wealth taxes are levied as percentages of the federal rates (Federal Tax Administration, 1991). It might be considered that the taxable capacity of private owners would be trivially small if similar to the net returns obtained from public forests. This is probably true for income, but not for wealth as forested lands still have value.

Income tax

In most cantons, the procedure for assessment for income tax requires that an average increment and an average roadside price be set by the cantonal authority, and that, on the basis of a points system reminiscent of a work study standard time table, appropriate harvesting costs are allowed. Cantonal forest staff are responsible for checking the site and crop factors on which the points are based, as they are for reviewing the silvicultural and protection expenses which are deductible. In recent years the outcome of such assessments is usually to yield a trivial tax. As a number of authorities consulted indicate (Amstutz, personal communication; Gussmann, personal communication), the tax has only a small impact, because the yield can be neglected. However, since there must be some owners who pay the tax, a notional average figure of 5 Sw.fr. is adopted here.

Property tax

This is based on the official land value. Assuming a typical value countrywide of 2000 Sw.fr. ha^{-1} and an average rate of 0.1%, the tax amounts to 2 Sw.fr. ha^{-1} $year^{-1}$.

Wealth tax

This tax is also based on the standard land value for the area in question. It is in effect the only tax to which the bulk of forest owners are subject. The rate of this tax for net wealth assessed at 250,000 Sw.fr. is *c.* 0.4% implying a charge on land assessed at 2000 Sw.fr. or 8 Sw.fr. ha^{-1} $year^{-1}$.

Inheritance tax

All cantons but one levy this tax, but 11 levy no tax on inheritances where the heir is a child (details are set out in Interkommunal Kommission für Steueraufklarung, 1991, a work whose title suggests that a definite need had to be met). Elsewhere rates on such transfers are low; because the burden is small it is neglected here.

Summary of direct tax charges

Broad average tax payments on a countrywide basis are summarized as follows (Sw.fr. ha^{-1} $year^{-1}$):

- Income tax 5
- Property tax 2
- Wealth tax 8
- Inheritance tax neg.
- Total 15

GRANTS

Substantial grants, in the form of a combination of Federal and cantonal funds, are, in principle, available for almost all forest operations apart from restocking after planned felling, and general maintenance tasks. In 1992, compared with the previous year, however, there was a 20–30% reduction in the sums available across all operations. The 1991 grant rates on which the following summaries have been based take account of the region, the operation and the owner's financial position in respect both of capital and income. This last element can account for a variation of 15 percentage points in the aid offered.

Afforestation

- 34–58% of the cost of new planting;
- 30–60% of weeding, cleaning, protection and access operations in young stands.

Restocking damaged stands

- 60–100% of the costs of removing storm damaged trees, the exact proportion depending on the logging cost level;
- 23–40% of the cost of restocking storm-damaged stands which require clearance.

Protection

- 75% for protection operations to counter damaging agencies such as insects.

Avalanche control works

- 40–75% for works for protection against avalanches, rock falls and land slips.

Total grant payments

An overall assessment of grants paid relative to the costs of works is given in Office Fédéral de la Statistique (1991). This shows the following proportions for different kinds of operation in 1988:

1. Planting, sowing, cleaning, torrent control, avalanche works, roads and cableways related to protective works, fences; grants 62.6m Sw.fr. relative to total costs of 100.8m Sw.fr., or 62%.
2. Other roads and cableways; grants 33.4m Sw.fr., costs 84.2m Sw.fr., or 40%.
3. Consolidation of holdings (roads, other works); grants 5.4m Sw.fr., costs 18m Sw.fr., or 30%.
4. Conservation works; grants 44m Sw.fr., cost 107.1m Sw.fr., 41%.
5. Grand total 145.4m Sw.fr. grants relative to costs of 310.1m Sw.fr., or 47%.

In addition, investment credits at zero or low rates of interest can be provided for approved building operations or the remaining (i.e. unaided) element of certain forest operations. In 1991 such credits amounted to 7m Sw.fr. compared with grants totalling 110m Sw.fr.

These amounts apply to cantonal forests as well as private. In the absence of direct information on grant payments by recipient, the ratio of areas in private hands to cantonal forest area, 28%, is adopted as the basis of allocation, to give $0.28 \times 152 = 43$m Sw.fr.

Extension

Extension is an important part of cantonal aid to owners. The organization has three levels: canton, region within the canton, and district. Expenditure on administration, management planning, supervision and advice amounted to 79m Sw.fr. in 1990 (Badan, personal communication). No separate figure is available for expenditure in relation to private forestry alone and the same basis as above, namely forest area, is adopted for allocation purposes yielding 0.28 × 79m Sw.fr., or 22m Sw.fr.

Research

The principal organization conducting research is the Federal Snow, Water and Forest Research Institute, Birmensdorf. Forestry accounted for an expenditure of some 29m Sw.Fr. in 1990. Research by the Forestry Department of the Eidgenössische Technische Hochschule (ETH) is estimated to amount to an additional 10m Sw.fr. (Schmidt-Haas, personal communication), and a further sum is directly financed by both the federal forestry authority and by cantonal departments. In total a value of some 40m Sw.fr. may be estimated but the allocation to private forests is problematical. Even allowing for the substantial extension activity of cantonal foresters, it seems improbable that research information passed to private owners is on a basis as high as that suggested by their area (28%) and a 22.5% allocation is therefore adopted, leading to a sum of 9m Sw.fr.

Summary of Public Expenditure in Support of Private Forestry

Total estimated expenditure on direct support of private forestry in recent years as follows (Sw.fr. $year^{-1}$):

- Grants 43m
- Extension services 22m
- Research 9m
- Total 74m

Conclusions on the Implementation of Policy

The most impressive aspect of the implementation methods employed is the willingness of all communities and authorities to bear the heavy costs noted above to support the continued management of forests. Tax burdens are not negligible, but grant rates are high, services in kind are valuable and it appears likely that harvesting will be subsidized in future. The forest is looked upon as a means of keeping Switzerland in business, and the management standards adopted appear to be in the tradition of 'business-as-usual'. The cost and efficiency implications of continuing to apply such a policy in the long term are important. Because the sums are not nearly as large as those involved in the support of employment and agriculture they do not excite the attention they might. However, in principle, they raise issues both of policy and of management reminiscent of the concerns in Britain over social provisions in other fields such as health and education. In that context it is perhaps surprising that public sector forestry is not subject to more rigorous financial disciplines.

Outstanding Issues

The issue of the location of compensatory planting appears to have attracted disproportionate attention in Switzerland, bearing in mind the fact that the annual area of forest clearance amounts only to 100–200 ha. But it must always be remembered that if the control of clearance were not operated through the system of compensatory planting, the area of clearance would increase significantly. Nevertheless, it is useful to follow through the arguments that arise in that country because they raise in acute form the fundamental questions that apply in relation to the current conventional wisdom on restocking in most developed countries, namely that clear felling shall be followed by restocking. The requirement that deforestation must be balanced by appropriate and appropriately sited compensatory planting is a less common obligation elsewhere but is a further important consideration in Switzerland.

The first point to consider in relation to an insistence on retaining the same total area of forest both in the country as a whole and broadly by region is a purely logical one: namely there is no reason why the exact disposition of forest existing at a particular time in all parts should be the 'best'. This is indeed recognized, since plans can be drawn for the afforestation of certain sites in order to match the particular future protection requirements of the locality by a change in land use to forest. However, decreases in the total area of forest are strongly resisted by foresters even in the plateau areas of Switzerland where it is argued that the removal of forest for other forms of land use creates dangers for existing settlements. Such a view appears extreme. True, forest clearance may alter the hydrological balance to some extent, thus raising run-off and allowing snowmelt to proceed more rapidly, but is doubtful whether such effects are usually more than trifling though there must of course be exceptions, as on steep valley sides. The real reasons for the attitude concern recreation needs in the vicinity of urban areas and the desire to control urban sprawl as discussed below.

Against these considerations of the desirability of entertaining change in the forests' distribution may be set these points. First, the principle of equivalent compensatory reforestation has maintained a balance which many would favour between built-over land and countryside. This has averted (according to one view) or blocked (according to another) the systematic colonization by building developers of former forest land in the Oberland and other populous areas. The financial incentives in favour of such change are naturally enormous, since such land use raises the value 500, maybe 1000, times (Badan, personal communication). Second, the measure has avoided speculative purchases of forest land. This consideration must, however, be respected as a significant factor only if the market is regarded as a totally misleading guide to action. Third, the principle of equivalence has led to planting areas in the neighbourhood on steep relief, alongside streams or as strips dividing built-up land. In other words it can be argued that the outcome has proved defensible in terms of sensible land use. There has, it is true, been a switch to more marginal lands which appears desirable on many grounds, and all this has been achieved without the bureaucracy of compensation payments. The recreational demands of an increasingly urban population have simultaneously been safeguarded relatively close to their homes.

The issues raised in Switzerland can be parallelled in most countries, with only minor changes in the considerations. Even where widespread afforestation is taking place, as in Britain, the subject of obligatory restocking, the 'politically correct' solution, originally favoured on timber conservation grounds and maintained since the 1960s on landscape grounds, deserves to be debated.

As already noted, the whole matter of the outlook for marketing and standing tree values is a critical one in Switzerland. There are two additional points to note. The first is that data on standing value are not obtainable in the country, they have instead to be calculated. This alone creates a substantial barrier to rational analysis of the position. Second, the costs of the silvicultural systems adopted and their component operations call for consideration if a determined attack on the uncompetitive nature of forestry is to be mounted.

A separate issue concerns hunting. It is noteworthy that hunting societies no longer attract young members; more significant still is the fact that hunting is banned in the Canton of Geneva.

References

Note: the Federal Ministry of Environment, Forestry and the Countryside = Bundesamt für Umwelt, Wald und Landschaft (BUWAL) = Office fédéral de l'environnement, des forêts et du paysage (OFEFP).

Büchel, M. (1991) [Forestry training : the new PROFOR programme.] In: *Annuaire Suisse de l'Économie, 1989*. Office fédéral de la statistique. Berne.
BUWAL (1991) Roundwood prices and specifications. (Response to FAO/ECE enquiry.) Typescript, Bern.
Chancellerie fédérale (1990) [*Forests*]. Office central fédéral des imprimes et du materiel, Berne.
Conseil des Etats (1988) [*Forest Law: Proposal of the Conseil Fédéral.*] Berne.
Conseil des Etats (1991) [*Federal law on forests.*] (Bill as of 4 October 1991.) Berne.
Conseil fédéral suisse (1988) [*Message Concerning the Federal Law on the Conservation of Forests and Protection against Natural Disasters.*] Berne.
ECE/FAO (1992) *The Forest Resources of the Temperate Zones*. Geneva.
Federal Tax Administration (1991) *Federal, Cantonal and Communal Taxes: an Outline of the Swiss System of Taxation*. Berne.
FAO (1988) Forestry policies in Europe: Switzerland. *FAO Forestry Paper* 86, Rome.
Federal Tax Administration (1991) *An Outline of the Swiss System of Taxation*. Berne.
Interkommunal Kommission für Steueraufklarung (1991) [*The Swiss Tax System: Tax Charges on Inheritance.*] Bern.
Office Fédéral de la Statistique (1991) *Annuaire Suisse de l'Économie Forestière et de l'Industrie du Bois*. Berne.
OFEFP (1989) [*The Swiss Forest Today.*] Berne.
Peter, D. (1989) [Management outturn 1988]. *Wald & Holz* no. 2, 120–137.
Prats Llaurado, J. (1991) Forestry administration. *Proceedings World Forestry Congress* 7, 261–272. Paris.
Price, M.F. (1988) A review of the development of legislation for Swiss mountain forests. *Arbeitsberichte des Fachbereichs*, no. 88/9. ETH, Zurich.
Siegfried, A. (tr. E. Fitzgerald) (1950) *Switzerland: a Democratic Way of Life*. Jonathan Cape, London.
Swiss Federal Office for Forestry (1987) *Mountain Forest*. Berne.
Swiss Federal Office for Forestry (Undated) *Forestry in Switzerland* (tr. M. Bigler). Berne.
World Bank (1990) *The World Bank Atlas*. Washington, DC.

12

DENMARK

BACKGROUND

Of the countries considered in this review, Denmark is one of the closest in many social, cultural and economic aspects to Britain, certainly in respect of the comparison with the British lowlands. It also provides a model of how silviculture and a wide range of objectives may be harmoniously combined under conditions in which pressures on the land and the requirements for environmental benefits of the land are high. The following statistics show the broad similarities between the two countries (values for Denmark quoted first).

1. Degree of urbanization lower as is density of population per hectare of land surface at: 1.2, UK 2.4.
2. Forest as proportion of land area: 10.3%, 9.9%.
3. Forest per head of population: 0.09 ha., 0.04 ha.
4. Proportion of forest owned privately: 73%, 57%.

Environmental concerns are marked in Danish land use, including worries over groundwater quality especially in the east of the country (cf. East Anglia), special protection of dunes from 'development', and a strong popular demand for more broadleaved trees in both existing forests and new plantations.

There are, however, major differences between the two countries because of their very different 19th century histories. Until the latter half of the 18th century, most of the land in Denmark was in the hands of the king and a small number of autocrats and the peasantry was heavily oppressed. The area of woodland shrank to a minimum in the early decades of the 19th century when both agriculture and forestry were at their nadir. Some improvement was achieved through agrarian reform undertaken in the last decades of the 18th century. Among the leaders in this process was Count Reventlow who is famed for his espousal of heavy thinning and for one of the first published expositions of the use of present value calculation as a guide to action in forestry (Grøn, 1960). The lack of timber, although an inconvenience, was not disastrous, since Norway was a Danish fiefdom and wood was readily obtainable from the south east of that country. It was, however, Denmark's misfortune to choose, or more strictly to be obliged to choose, the wrong side in the Napoleonic Wars. As a result, the Danish Navy suffered a substantial defeat by Nelson in 1801. Later, in 1807, after much damage to Copenhagen, Denmark was to suffer the ignominy of having its fleet taken by the British. Depleted by the continental blockade, relieved of Norway in 1814, the country found it impossible to sell grain to Britain because of the protection given by the Corn Laws. In mid-century, the potential for a renewed grain trade to Britain and Germany rose but prices fell with the opening up of Russian and North American supplies. However, improvements in internal transport and in agricultural techniques led to the growth of bacon and butter production and the growth of substantial export markets. A feature of this phase of agricultural growth was that farmers organized their exports on a cooperative basis. The reliance on cooperative effort is characteristically Danish: it is echoed today in the success of organizations for wood sales and forest management.

At the same time there was a steady growth in the area of forest, rising from its low of 150,000 ha in the 1820s to almost double that figure by 1900. The State was very active in raising new forests, mainly in the western province of Jutland as early as the late 1700s, and afforestation continued during the 19th century both on marginal agricultural land in that province and on large estates in other parts of the country (Plum and Honoré ,1988). The real feature distinguishing Danish from British forestry is the longer and far stronger commitment to forestry as a form of economic land use and the success of landowners in making their enterprises profitable. Because of the different agrarian histories of the two countries, this applies as much to owners of small woods as to large estates; an excellent introduction to forestry practice with special reference to small woods illustrates the continuing and generally rewarding attention given to the subject (Holmsgaard, 1981). In more recent times Danish foresters have compounded their success by becoming Europe's largest producer of Christmas trees and greenery with an export value in 1990 of over 400m D.kr.

Today private forests are in the hands of individuals and various foundations and companies. In the main, the smaller properties are owned by farmers, some 220,000 woodlands forming parts of farms and accounting in total for one-fifth of the country's total forest area (Helles, 1988). Two-thirds of the area lies in holdings of over 50 ha, usually forming part of estates (Dansk Skovforening, 1982). This broad spectrum of woodland owners has created its own stimulus for successful cooperative measures as outlined below. An important source of income is the sale of Christmas trees and greenery.

Policy and Legislation

Until 1989, Danish forestry and the systems supporting the private sector had been steered by the Forest Act of 1935. That law followed the tradition of the Forest Act of 1805, which laid emphasis, to the exclusion of any other consideration, on the provision of a large and valuable (*sic*) wood supply. The growing demand for recreation and elements of it, or running alongside it, such as nature conservation, amenity and protection of features of cultural heritage, led to a number of studies in the 1980s. These culminated in the Forest Act 1989 and the Nature Management Act of the same year (Anon., 1991a, b). These laws epitomize the country's forestry policy and the intended role of private owners in it. Both acts emerged after much discussion and consultation and won almost complete support from all parties in their passage through Parliament. A theme that has been developed in recent years is 'people, trees, woods' and the multiple use idea of this is basic to the laws and a continuing part of associated educational programmes. Laws mentioned here and ones yet to be enacted form part of a major legislative programme termed the 'green law package'.

Forest policy can be identified from Section 1 of the Forest Act:

> 1(1) The purposes of the Act shall be:
> 1) to conserve and protect Danish forests,
> 2) to improve the stability, structure of ownership and productivity of forestry,
> 3) to contribute to increasing the total forest area, and
> 4) to strengthen advisory and information activities.
>
> 1(2) In administration of the Act importance shall be attached to ensuring that forests are managed in order to increase and improve wood production and to protect landscape amenity, nature conservation, cultural heritage and environmental-protection interests, as well as recreational activity interests.

The Nature Management Act of the same year reinforces certain of the forestry aims. Thus Section 1 states as purposes:

> 1) to conserve and tend landscape amenity and cultural heritage values,
> 2) to conserve or improve the conditions for wild plant and animal life,
> 3) to increase the forest area, and
> 4) to improve the general public's opportunities for recreational activities.

The principal objectives are noteworthy. The first aim of the Forest Act follows the long tradition from the last century of ensuring that the use of forest land as forest should not normally be allowed to change. Since 1805 there has been an obligation placed on the vast majority of Danish woodlands that these should remain forest. The rule has been followed strictly; if conversion is permitted compensatory afforestation must be carried out. Forest reserves are legally constituted to ensure the protection of the area as forest. The splitting of such obligatory forest into lots of less than 50 ha has been prohibited since 1819: today the area of such small ownerships amounts to only 25,000 ha.

'Good' and, linked with that word, 'multiple use', forest management is defined in the Act. Thus planting should follow rapidly after felling if natural regeneration or sowing is not to be adopted, thinning should be favoured, and final felling, which should not take place before 'maturity', should favour the creation of a stable and varied forest. Management for Christmas trees and greenery from short rotation crops is not to exceed 10% of the forest's area, the edges of forests should be conserved with, preferably, broadleaved trees and shrubs, oak scrub will be conserved as will wetlands, bogs and heaths.

Particular attention is paid in the Act to the enhancement of broadleaved woodland. This implies the establishment of forest edges, a particularly Danish – and attractive – feature, using broadleaves, the maintenance of the area of oak and beech, production of suitable broadleaved seed and plant material, etc.

The afforestation of land must be accepted as part of a regional land use plan. Though the Act does not state the quantitative goal for new planting, the current government plan is to achieve a doubling of the total forest area at a steady rate over 70–80 years. Once afforested with the aid of grants, the area becomes obligatory forest.

It is interesting to note that, as in other countries with highly fragmented woodlands in different ownerships, the Minister is empowered to promote the amalgamation or joint management of forests. Past efforts in these directions do not appear to have been successful and in practice the matter is not regarded as an important policy issue. Also the Minister can require an owner to supply accounting information for his property. The Minister is advised by a newly established 13-person Forestry Council, which is widely representative of countryside interests, and is expected to have substantial influence. A recent task of the Council has been to comment on the new broadleaved woodland scheme.

Administration

In 1971, Denmark was the first country to create a Ministry of the Environment. Perhaps most clearly after the trauma of the Napoleonic Wars and the loss of Norway, but also on many subsequent occasions, Danes have demonstrated their ability to act as leaders of social change. Examples are provided by the development of the Folk High School movement from 1844 and the abolition of leasehold tenure of land (Oakley, 1972), and, in more recent times their open-minded attitude to freedom of expression in the form of pornography. The same willingness to be in the vanguard of change is evidenced in the administration of forestry, the responsibility for a major part of which since 1975 has been that of the environment ministry. Another 'first' has been the merger in 1987 of the nature conservation agency with the state forest service to form the National Forest and Nature Agency. But by no means all responsibility for forestry policy or its implementation is in Environment's hands. It is always difficult to gain advantages in one direction by a particular allocation of tasks to ministries without creating further problems. The promotion of afforestation of agricultural land remains the responsibility of the Ministry of Agriculture, as does the grant aiding of a wide range of improvement operations and the support of the principal consulting/contracting bodies for private forestry noted below. In addition hunting remains an agriculture subject. Furthermore, the various state research institutes concerned with trees, forestry and forest operations in Denmark are scattered among the Ministries of Agriculture, Environment and Education, with the principal body, the Danish Forest and Landscape Research Institute, under the Minister of Agriculture.

No discussion of Danish forestry would be complete without reference to the three owners' organizations. The largest in terms of staff is the Danish Land Development Service (DLDS), better known to many British foresters as the Danish Heath Society (Hedeselskabet). This was founded in 1866, undertakes contracting services and carries out extension work. Next largest in terms of staff and turnover is the group Danish Forest Owners Cooperatives (DFOC) (De Danske Skovdyrkerforeninger). This group of trading bodies carries out contract work for members and others and organizes sales (De Danske Skovdyrkerforeninger, 1990).

The Danish Forest Society (Dansk Skovforening) is the country's forestry society acting in the interest of forestry as a whole (Thomsen, personal communication). It produces a journal (*Dansk Skovbrugs Tidsskrift*-DST) in support of its professional scientific function; in this sense it is comparable to the Institute of Chartered Foresters with its journal *Forestry*. It also has as a principal responsibility of representing the interests of owners of larger forest properties and in that connection produces the journal *Skoven*. A very effective marketing section deals principally with sales of wood overseas.

None of the Danish cooperative organizations require that members sell all their wood through the particular Association or Society.

Implementation

Land use planning is conducted by the provinces. By an amendment of the planning legislation (Bjørnskov,1989), the local authority will decide areas in which afforestation will be favoured, areas where it will not and indeterminate areas. The total area available in the first category easily exceeds the government target. The considerations underlying the definition of areas in which afforestation will be favoured and the ways in which guidelines should be set up are described in Miljøministeriet (1990), and illustrations of such considerations as layout and design are given in Pedersen (1990).

Fears have been expressed by landowners, both farmers and foresters, that the current dependence on the state in undertaking this programme heralds expropriation. The lack of enthusiasm is dramatically illustrated by the fact that the rules for grant-aiding this activity were set up in 1989, but in 1991 only three enquiries for the grant were received by the Ministry of Agriculture. Indeed the Danish Forestry Society (Dansk Skovforening, 1991a) suggested that for unstated and unexplained reasons there was a political unwillingnes to favour private afforestation and the income tax block was employed as the last bastion to sustain this position. However, uncertainties over the virtues of afforestation were somewhat reduced by a new Act of 21 December 1991 which made afforestation expenditure deductible against income.

TAX

It is clear from the sample returns compiled by the Danish Forestry Society (Dansk Skovforening, 1991b) covering the over-50 ha category in so-called 'old' forest and heath forests, that net income and capital values of forest land and associated assets differ greatly across size-classes of property and location. The calculations of tax burdens set out below have been based on figures for a countrywide average for these larger estates, two-thirds of which are assumed to be those termed 'old' woods and one-third heath forests.

Property tax

There is a tax on property (grundskyld). Its burden is not negligible but it is deductible against income. The burden is assessed at an average of 160 D.kr. ha^{-1}.

Income tax

In established forest, silvicultural costs are immediately deductible against income from any source. This applies to the production of greenery as much as to trees grown for wood. Thus restocking, as in all other countries considered in this study which have income taxation of forestry, is tax assisted. However, while road maintenance is deductible, expenditure on new roads is not. This naturally leads to some creative use of the term 'maintenance' in relation to improvement operations.

Income tax has two elements, one on income derived from capital assets such as forest, the other on labour income. The scale of each is highly progressive: for the owner of an average-sized woodland estate an average rate of 50% is assumed. A countrywide average net income, after deduction of property tax, is assessed at 600 D.kr. ha^{-1} $year^{-1}$. The burden of income tax is thus calculated as $0.5 \times 600 = 300$ D.kr. ha^{-1}.

Wealth tax

This tax is payable on net assets of over 1.3m D.kr. at a rate of 1.5%. Relief of 80% is provided on forest and other business assets, so that the rate becomes 0.3%. In addition, below the total tax ceiling which is applied, further relief is available of 60% reducing the effective rate to 0.12% (Holten-Andersen, 1985). As is common with taxes on wealth, the room for manoeuvre in portfolio management is large and it is believed that taxpayers are efficient at avoiding much of the tax through appropriate management so minimizing their net taxable wealth (Holten-Andersen, personal communication). One route may be to purchase forests by means of a loan, so reducing net assets. Such activity causes speculation, as well as that *bête noire* of forest managers, discontinuity of management. Although at the lowest level an owner might evade wealth tax, the assumption made here is that for an average property in one person's continuous ownership a marginal rate of 0.12% applies, implying an average rate taking into account the zero rated element of net assets of 0.1%. With an average forest value of the order of 30,000 D.kr. ha^{-1}, wealth tax thus amounts to 0.1% × 30,000 or 30 D.kr. ha^{-1} $year^{-1}$.

Inheritance tax

Inheritance tax is regarded by forest owners, and with justice, as the major tax problem. It is levied on transfers at death to a child or grandchild at rates of 10% up to 1m D.kr. and of 32% over 1m D.kr. No special relief for forests is provided though an allowance of 15% on business assets also applies to forest property. For a transfer of 5m D.kr. the average rate of tax is 27.6%. It would be naive, however, to assume that beneficiaries of inheritances suffer the substantial charges that follow from the combination of moderate rates and high and barely relieved values. Rearrangement of the ownership of assets of all forms as elsewhere diminishes the impact of the tax, as does generation skipping. Assuming an effective average rate of 15%

on a value of 30,000 D.kr. ha^{-1} relieved to the extent of 15% and an average interval of 30 years implies an equivalent of $0.0246 \times 0.15 \times 0.85 \times 30{,}000$ or 94 D.kr. ha^{-1} $year^{-1}$.

Summary of direct tax charges

Total tax payments on a countrywide average basis are thus estimated as follows (D.kr. ha^{-1} $year^{-1}$):

- Property tax 160
- Income tax (after deduction of property tax) 300
- Wealth tax 30
- Inheritance tax 94
- Total 584

The total figure calculated for tax is nearly 77% of net income before tax. It should be noted that recent levels of cut and hence of income have been below the long-term average (Dansk Skovfarening, 1991b, Figure 2). Even so, the above figures are higher than those of other countries reviewed in the present study. One factor influencing the situation in Denmark may be the contribution which greenery and Christmas trees make to sales and to net income. However, the major cause lies in the tax system itself which makes only small allowance for forestry as a special activity.

Grants

The Ministry of Agriculture is responsible for all the current grant schemes for forestry and tree planting except the broadleaved planting scheme in obligatory forest which is the responsibility of the Forest and Nature Agency. Disbursements by the Ministry of Agriculture are accordingly much larger than those of the Forest and Nature Agency. The main lines of the various schemes are set out in Landbrugsministeriet (1989, 1990, 1991) and Skov- og Naturstyrelsen (1991a).

Afforestation, including planting of setaside land

Rates differ between areas where afforestation is favoured and those where it is simply permitted:

- Broadleaves: favoured areas 75%, permitted areas 50%;
- Conifers: favoured areas 60%, permitted areas 40%.

Grants are only available for areas of over 2 ha, smaller areas being considered for grant aid under the improvement scheme described below where the planting is linked to existing obligatory woodland. The maximum grant allowable is 30,000 D.kr. ha^{-1} for broadleaved planting and 15,000 D.kr. for conifers.

Regeneration of obligatory forest

1. The grant rate for conifers is zero.
2. The rates for broadleaves, in D.kr. ha^{-1}, are shown in Table 12.1.
3. Where the new stand is enriched with the addition of other native species characteristic of the locality at a density of not less than 50 trees ha^{-1}, the rates are raised by 10%. The qualifying species are: hornbeam, birches, black alder, wych elm, gean, rowan, small-leaved lime, Norway maple, field maple, sallow and crab apple.

Table 12.1. The grant rates for broadleaves (D.kr. ha^{-1}).

	Planting	Natural regeneration	Sowing
Beech	15,000	9,000	6,000
Oak	10,000	9,000	6,000
Ash	8,000	8,000	6,000

Note: Grants are payable in two instalments, the second a maximum of 8 years after the first in the case of afforestation or sowing, and a maximum of 16 years in the case of natural regeneration.

Shelter belts

1. Collective planting by not fewer than 20 farmers and involving not less than 20 km of belts, carried out under a joint plan and by a single contractor may be grant aided to the extent of 50% of expenses including maintenance to the end of the third year.
2. For individual planting carried out by the farmer himself the grant is 50% of the cost of the plants alone.

Improvement

1. Grant aid is available for a wide range of operations, excluding the raising of Christmas trees and greenery, which are in conformity with the guidelines for multipurpose forestry: all woodlands over 0.5 ha are admissible. Operations include cleaning, first thinning, enrichment of low production stands, broadleaved belts for fire or wind protection, roading, draining and drain cleaning, preparation of management plans. Rates are 50%, except for belts and planning for which the rates rise to 60%.

2. As originally devised in 1989, the scheme offered declining rates of grant for properties of between 50 and 150 ha with no grants for larger forests. Uptake of the budgeted provisions has, however, been low at 25% and more liberal aid has been paid.

Extension

Public expenditure on this activity arises on two fronts. First, forest districts in the Forest and Nature Agency are responsible for the administration of the various schemes financed by the Agency and for certain extension services. Second, the Danish Land Development Service and the Danish Forest Owners' Cooperatives receive grants for the consultancy work they do for the benefit of owners of small forests. These grants, which are paid by the Ministry of Agriculture, are on a graduated scale varying from 50% of the salary, etc. and travel expenses of the consultant for woodlands under 50 ha, with no provision for forests over 150 ha in the more fertile regions of the country and 30 ha in the heathland region. In 1991 total outgoings were 13.8m D.kr. to which may be added some allowance for Forest and Nature Agency spending making a total of some 15m. D.kr.

Research

The State Forestry Research Institute (Statens Forstlige Forsøgsvaesen), the Institute for Forest Technology (Skovteknisk Institut) and two minor institutes were merged to form the Danish Forest and Landscape Research Institute (Forskningscentret for Skov og Landskab) in 1991. Together with expenditure of the Royal Veterinary and Agricultural University, total spending on research amounted to some 50m D.kr. in 1990 (Elers Koch, personal communication). However, 18m D.kr. of this sum was spent on hunting and landscape research, implying 32m D.kr. on conventional forest research. Of the total 20% is privately financed and assuming an allocation to private forestry of 75% of the residual, the outcome is 19m D.kr.

Summary of Public Expenditure in Support of Private Forestry

Total public expenditure in 1990–1991 attributable to private forestry was as follows (in m D.kr.):

• Grants: Forest & Nature (Ministry of Agriculture)	
Broadleaves	6.6
Improvement	6.3
Shelterbelts	23.0
• Extension	15.0
• Research	19.0
• Total	69.9
(excluding shelterbelts	46.9)

Comments on Policy Implementation

It is clear that the change in responsibility for state forestry and part only of private forestry leaves a good deal to be desired in terms of the administration of forestry matters both within government itself and in relation to forest owners and farmers. That some rationalization is desirable is clear but the outcome must depend on political decision.

Forestry taxation excites much attention in Denmark. In every country problems arise in attempting to square the circle of avoiding too discriminatory a regime for forestry activities with the provision of an appropriate level of incentives to achieve society's aims, a proportion of these coming from grants. This task is more difficult the higher the proportion of revenue raised by direct as opposed to indirect taxation and where, as in Denmark, social security payments are financed largely through direct taxation. The position is also complicated by owners' unwillingness in many cases to accept manipulation through the medium of grant systems and rates that may be varied more readily than taxes. Support of cooperative movements among owners is strong and the organizations' role is valued by government. Further study of their structure and performance is justified.

Outstanding Issues

A principal issue concerns the choice of the mix of activities that a woodland owner may adopt in his forest in the light of the clear change of emphasis to multiple use forestry. Whereas owners are willing to undertake the switch to a more diverse pattern of forestry, the technical prescriptions are sometimes lacking. This places a responsibility, recognized in the Forest Act, on the Forest and Nature Agency to lead by example as well as to provide a substantially greater information and advice service than in the past. As Bjørnskov (1989) points out, the desire is to place less reliance on legal means of achieving ends and more on support measures. In part this solution is forced on government simply because it is increasingly impracticable to rely on regulation and definition to encourage owners to decide on appropriate action in the light of the varied emphases that may properly be given to the different objectives and of the variety of operations and situations that exist (Skov- og Naturstryelsen, 1991b).

The proposal for private owners and the state combined to afforest 5000 ha of agricultural land annually creates two problems. The first is that planting with the aid of grants means that the owner accepts that the land has acquired the status of 'obligatory' forest. In Denmark, this familiar feature carries with it such resistance to the acceptance of the implied restrictions on freedom of action that the value of the land may be reduced (Thomsen, personal communication). An interesting enquiry mounted by the Agricultural University for the Ministry of Environment (Jensen, 1989) gives some support to this idea. A set of options was presented to farmers setting grants and subsidies against the acceptance of certain obligations. The obligation to preserve the forest in perpetuity considerably reduced the appeal of afforestation. While the conditionality of grants which maintain the farm–forest link must reduce the flexibility of sale, it appears that in practice estate values are enhanced where bare land is planted (Raae, personal communication). Similar conflicting outcomes can be seen in Sweden and Britain; the point is that value depends on what, realistically, the alternatives are. Alongside this concern about changes in property value is the notion that foreigners may buy land for afforestation. A second issue concerns the usual underlying problem of enticing private investors to undertake the rate, kind and even location of investment that society desires.

Public access to private forests for walking and bicycling is allowed in daylight hours where there are access roads to the area. This right is balanced by an arrangement for government money to be paid as compensation if damage is done. In practice, few cases of damage are brought up for such treatment.

Hunting rouses passions on both sides. Although landowners have had the right since 1971 to close areas and even rights of way for shooting, this right has been wisely exercised. However, it clearly tends to be thought of as inconsistent with general recreation in forests. In addition there is support for the feeling that shooting is necessarily out of sympathy with nature conservation. Discussion of the basis of a change in the law began in 1991; a new law is expected to be passed in 1993.

References

Anon. (1991a) The Forest Act, 1989. (Engl translation.) Miljφministeriet, Copenhagen. Typescript.

Anon. (1991b) The Nature Management Act. (Engl translation.) Miljφministeriet, Copenhagen. Typescript.

Bjørnskov, L. (1989) Address to the (British) Countryside Recreation Research Advisory Group Conference 'People, trees and woods '. National Forest and Nature Agency, Hørsholm. Typescript.

Dansk Skovforening (1982) *Brief Glimpses of Danish Forestry.* Copenhagen.

Dansk Skovforening (1991a) [*Annual Report: Forestry and Growing Industry.*] Copenhagen.

Dansk Skovforening (1991b) [*Accounting Results for Danish Private Forests 1990.*] Copenhagen.

De Danske Skovdyrkerforeninger (1990) [*The Danish Forest Owners' Cooperatives 1940–1990.*] Copenhagen.

Grøn, A. H (ed.) (1960) *A Treatise on Forestry by Count C.D.F.Reventlow.* Society of Forest History, Hørsholm.

Helles, F. (1988) Denmark. In: Forestry in the EEC during the Eighties. *FAST Report* no. 161. CEC, Brussels.

Holmsgaard, E. (ed.) (1981) [*Management of Small Woods and Plantations.*] Landhusholdningsselskabets Forlag, Copenhagen.

Holten-Andersen, P. (1985) [The Danish income and wealth tax system illustrated using three-dimensional graphs.] *Dansk Skovforenings Tidsskrift* 70, 206–218.

Jensen, F.S. (1989) The farmer and the marginal land. *Geografisk Tidsskrift* 89, 4–48.

Landbrugsministeriet, Copenhagen (1989) [*Grants for Forest Improvement.*] (1990) [*Grants for Shelter Belt Planting.*] (1991) [*Grants for Establishment of Forest on Agricultural Land.*]

Miljøministeriet, Planstyrelsen (1990) [Afforestation and where planting is not favoured.] *Dansk Skogbrugs Tidsskrift* 75 (3), 88–107.

Oakley, S. (1972) *The Story of Denmark.* Faber, London.

Pedersen, O. (ed.) (1990) [Special issue on afforestation.] *Dansk Skovbrugs Tidsskrift* 75(3), 81–185.

Plum, P.M. and Honoré, B. (1988) Forest policy and legislation. *Environmental Policy and Law* 18(4), 111–115.

Skov- og Naturstyrelsen *et al.* (1991a) [*Grants for Broadleaves.*] Skov-info Nr.3. Skov- og Naturstyrelsen, Hørsholm.

Skov- og Naturstyrelsen *et al.* (1991b) [*Multiple use Forestry – a Checklist.*] Skov- og Naturstyrelsen, Hørsholm.

13

NORWAY

BACKGROUND

Certain features of Norway set the country apart from both its nearest neighbours and other societies in Western Europe and, effectively, the rest of the world. The most prominent is the egalitarian attitude which pervades every walk of life as it has done for centuries (the nobility was abolished in 1821). The second is the jealous guarding of individual freedom, epitomized in the axiom 'Freedom with responsibility' (literally translated 'Freedom under responsibility'). But this desire for independence is not such as to stifle joint enterprise; enlightened self-interest finds no difficulty in embracing the cooperative spirit which is strong and especially so in agriculture and forestry. Third, close links with the land are maintained in policy and practice, urbanization and industrialization being relatively recent and the country still being heavily dependent on the exploitation of natural resources for its livelihood. Fourth, following from this background, and of significance to forestry in particular, there is a strong regional policy. Indeed in certain regards it is difficult to know whether a given measure is a forestry one or one serving settlement policy.

Five statistics are highly relevant to the following discussion.

1. Norway has the lowest population density of any European country, apart from Iceland, with 7.2 ha per person compared with 1 in France and 0.4 in Britain.
2. GDP per caput at $21,100 in 1989 (World Bank, 1990) is fourth highest in the world and equal to that of the United States.
3. 60% of the land area is fell country above the tree line.
4. The forest area per head of the population at 2.0 ha is relatively high, being 10 times the EC average.
5. Private individuals own 5.1m ha or 77% of productive forest land.
6. The average size of forest holdings over 2.5 ha is 43 ha.

The importance attached to land, its pattern of ownership, occupation and use is a distinctive feature of Norway and is a reminder of the country's relatively recent emergence from a rural society based on farming and fishing. There is a strong tradition in favour of keeping farm or farm/forest holdings within the family, a tradition which is enshrined in the Allodial Act (Odelsloven). This dates from more than 1000 years ago and has its roots in the protection of the farmer from the king and nobles. It establishes the rights of inheritance of landed property, 75% of all transfers of such property being between family members.

An important law, the 'Concession Act', is more recent; it governs transfers of ownership of agricultural and forest land outside the family. In 1905 the union of Norway with Sweden was dissolved. The starting point for the decision to legislate on land was the fact that a large number of waterfalls had been bought by Norwegian or foreign companies. In 1906, three-quarters of all developed water power was owned by foreign firms. Concerned that so many of the country's natural resources were falling into foreign hands or were liable to be monopolized by a small number of capitalists, Parliament passed the Concession Act in 1909. The splitting of properties is in principle not allowed, and although the most recent survey (1989) shows an increase in the number of land holdings, much of this increase arises because of improved coverage. It is not an unusual feature of the law in many countries to find a ban on ownership of land by foreigners but the restrictions placed on absentee ownership in Norway are noteworthy. Since 1974, both the Concession Act and the Allodial Act restrict the possibility of absentee ownership and limit the price level directly (Bjorå, personal communication). The Concession Act also gives the County Board the right of first refusal in order to make it feasible to assign the land to a local buyer, often a neighbour. In practice, this arrangement is not exploited at all frequently. The law, however, must stultify the development of the range of enterprise that might otherwise be found, for example by making the purchase and subsequent enlargement of properties except by close neighbours difficult. There are those who continue to argue in favour of maintaining the stability of rural communities and the associated culture that such a measure ensures, but it seems likely that different attitudes will dominate as the economic link with the farming way of life and indeed with a land-based livelihood fades. This subject interacts with regional policy, a topic referred to under 'Outstanding issues'.

Whether these social policies have helped or hindered the development of forestry is a moot point. Other factors have dominated events in forestry since the Second World War. Thus there have been a number of significant policy initiatives favouring direct investment in enhancement of growing stock value through roading, draining, cleaning, etc. In addition, low real rates of interest have encouraged borrowing from the capital market rather than growing stock, and the replacement (in 1954) of income taxation based on increment with one based on cash has reduced the pressure to cut. There has also been the traditional farmers' practice, for farmers are important owners of forest, that forest should not be cut heavily except in times of emergency. The result of these influences has been a major increase in the value of the forest stock; at the same time a continued rise in production has been attained. The growing stock increased from some 300m m^3 in 1920 to 575m m^3 by the early 1980s. Total fellings, including fuelwood, at 10m m^3 have remained well below the increment of about 17m m^3 (Hamilton, undated) and the theoretical sustainable level of yield of about 13m m^3 (FAO, 1988, p. 181).

Owners' associations play a major role in Norwegian forestry. Their functions range widely, with the central organization negotiating pulpwood prices with industry and promoting policies aimed at benefitting its members, and with the local associations negotiating sawlog prices and arranging sales of wood from members' forests, undertaking final fellings, making inventories and preparing management plans, and organizing training and education. The principal body, the Norwegian Forest Owners' Federation, owns 37% of the capital of the country's largest wood processing firm which processes nearly one-half of all home-produced industrial roundwood. The area of forest held by the wood-processing industry is significantly low at 450,000 ha, or 9% of private holdings.

LEGISLATION

The Norwegian tradition on the need to enshrine new policy departures in law is similar to Britain's, namely only major departures require legislation.

The last major forestry law was that of 1965: this Act was amended subsequently, notably in 1976, to incorporate clear environmental requirements. Its purpose as set out in Section 1 (Anon., 1988) was to 'promote forest production, afforestation and forestry protection'. The forest owner was given the right to mark trees for cutting so long as the silvicultural rules of the Act were obeyed. These state, among other matters, that thinning or regeneration felling must be such as to further favourable development of the forest from the point of view of quantity and quality. However, regeneration felling does not require permission, but restocking by whatever route is compulsory, as is the stocking of bare or partly stocked forest which has carried a stand within the previous 20 years. But the Act prescribes few compulsory measures regarding silvicultural practice and these few are seldom used. The Act requires the owner to have regard to recreation and the natural environment in relation to all operations. The authorities can require an owner to draw up a working plan where there have been certain deficiencies in management, and although the Act does not make the preparation of management plans obligatory for private owners, plans are usual. This is in accordance with the doctrine that the owner should be left free to manage his forest without intervention by the authorities so long as good practice is followed. Currently the accepted programme is that plans should be revised not less frequently than every 15 years; the current rate of preparation or revision is 400,000 ha annually.

The Act also confirmed arrangements for the administration of the Forest Trust Fund first established as a legal requirement in 1932 (Anon., 1990). The Norwegian Forestry Society (1991) describes the Fund's function thus:

> In order to secure sufficient investment in the forest, the Forestry Act provides that the individual owner shall set aside a certain percentage of the gross value of all timber sales. This fund is called the Forest Trust Fund. The owner may decide the percentage within the range of 5–25%. The Forest Trust Fund is deposited in a special account, which may be used to cover expenses for various silvicultural activities, construction of forest roads and other necessary undertakings in the forest from which the timber was cut. When using this fund the forest owner benefits from certain tax advantages.

The latter are important and are discussed below under income tax. It should be noted that the trust fund money goes with the property if a transfer occurs.

A 1989 report concentrated attention on environmental aspects; it led, *inter alia*, to the abolition of grants for operations involving the use of herbicides. A review of agricultural policy was undertaken in 1991 in the light of agricultural surplus production and the possibility of policy impacts of, or deriving from, becoming a member of the European Community. This review was undertaken by a government appointed committee (Alstadheim *et al.*, 1991), which also considered forestry because of the interdependence of the two sectors, as 65% of farmers are also forest owners (a higher proportion still, 90%, of forest owners according to Wefald *et al.*, 1991, also have farming interests). The committee's constitution is of interest: it was made up of 'qualified persons'. These included clearly interested parties such as representatives of forest owners as well as independent members. The outcome of the debate that this report will generate will no doubt be highly significant for the future of Norwegian forestry because of recent changes in attitudes to food production and to the needs of nature conservation. Some of the committee's points are noted below under 'Outstanding issues'.

POLICY

There is no one single source setting out Norway's forest policy, but the Minister of Agriculture has recently confirmed its main lines. These are founded on the laws mentioned above, on the report to parliament (Anon., 1985) and discussions on it in the Storting, and endorsement of the views of the Alstadheim committee (1991). This committee emphasized the need for forest policy to take account of the changed market position of farming relative to forestry and the long turnover period in forestry. The government favours a continuation of the present policy with its commitment to the natural environment and multiple use, and the support of regional employment and the supply of wood to industry.

The attitude of the forestry profession is to ensure a deliberate approach to change: policy development should, it is claimed, be a continuous process. The claim that policy develops in a continuous line is not without foundation or significance. No sudden departure on to a new track is favoured or, it is said, is found necessary. This view has the ring of a statement coming from a conservative group anxious to maintain the *status quo ante*, but there is good reason for believing it has force. The first is the argument from economics. Norway still obtains a significant proportion of its national income from forestry and wood processing, and from particularly disadvantaged regions. The foreign exchange earnings of the sector never cease to be paraded although at only 3–4% of the total (but 10% of non-oil exports) they are not huge. The second reason which is important and, as already noted, qualifies the income point, is that a strong concern underlying national policy in all sectors is the maintenance of the existing location of people and settlements. In a country with many natural drawbacks, notably of topography and shape – Norway spans 13° of latitude – this desire takes on a special meaning. Forestry with its extensive distribution naturally helps to sustain this desired arrangement.

In 1985, the agriculture department produced a report (Anon., 1985) in response to a government request for a review of one major aspect of forestry policy, namely the level of wood production that could be sustained. It is of interest to note the importance which the report attached to the potential of forestry to generate employment, especially in economically disadvantaged areas. But as in other Western countries, the same concerns over environmental aspects of forestry, and especially the manifestations of a highly mechanized and generally 'rationalized' form of forestry, have come to influence forest policy in Norway. A clear example of the impact of public concern was the preparation in 1984 of an official statement recognizing the area around Oslo (Oslomarka) as falling within the terms of the 1965 Forestry Act as a district requiring special treatment in order to satisfy multiple use, especially recreational, demands (Anon., 1985; Opheim, 1991). A further reflection of these pressures was the publication by the Ministry of Agriculture of the report of an official working party on multifunctional forestry (Aalde *et al.*, 1989). Today, grant assistance towards the preparation or revision of management plans requires that multiple use goals be taken into account in the prescriptions. An additional feature of the debate in Norway has concerned the contribution to multiple-use forestry made by the owners of the largest estates some of whom also possess major wood-processing mills.

Administration

The Forest Authority consists of the Minister of Agriculture, the County Land Boards and the Municipal Land Boards. Since 1981, the Land Boards at the two levels - fylke (or county) and kommun (district or municipality) - combine the functions of administering both forestry and agriculture. This integration with agricultural administrators is noteworthy, reflecting the close interconnection between the two activities. Members of the county land boards are elected by the county council on a political, proportional representational basis. The county chief forester is the secretary of the board in forest service matters. A similar arrangement applies at municipality level.

The Forest Service is the Authority's executive arm. Jointly with the Authority it has the following principal objectives:

1. To ensure that forest land is treated in accordance with the Forestry Act.
2. To administer incentives made available to forestry.
3. To provide professional advice and extension services to forest owners.

The Service's headquarters staff is small, numbering only some 25 persons, and total employment amounts to 500. It should be noted that the state forestry enterprise is quite distinct, being administered by a different department of the Ministry of Agriculture.

Close working relations are maintained with the two representative owners' associations. These are: the Norwegian Forest Owners' Federation (Norsk Skogeierforbund) with 56,600 members who own 3.7m ha and account for 75% of the country's total annual cut; and the Norwegian Forestry Association (Norsk Skogbruksforeningen) to which some 250 owners of larger properties belong, and accounts for about 12% of the total cut of the country. Formerly, working plans were prepared by the Management Planning Divison of the Norwegian Forestry Society, but since 1960 most management plans have been made by the owners' associations. A strong voice in favour of nature conservation is provided by the Norwegian Nature Conservation Society (Norgesnaturvårnforbund).

Higher education and research in forestry are administered by separate divisions of the Ministry of Agriculture and lower level education is administered by the Ministry of Education, Research and Church Affairs. The Forest Extension Service Institute is a partnership with 30 forestry organizations and scientific institutions as members. Its functions are to produce training material for local courses given by its partners, and to arrange in-service training to forest managers, owners and workers. Its eight permanent staff, assisted by specialists as required, are financed by fees and government grants.

Tax

The national system of income taxation changed on 1 January 1992. The shift involved two principal changes: reduced tax rates combined with a broadened tax base. For ordinary employment income, rates had been falling for some time: in 1986 the maximum marginal rate was 73.8%, in 1991 before the reform 57.8% and from 1992 48.8% (Eid, 1991). The shift in the tax base implied a greater dependence on actual income in place of calculated income. The change introduced a reversal in the distribution of maximum marginal rates between those on income derived from capital and income derived from labour. The former was effectively 40.5% before the change and 28% after, the latter was altered to 28% + 20.8%, or 48.8%. It should be explained that Norway has long operated two wealth taxes, that for payment to the State being on a sliding scale while that for local authorities was and remains at a fixed rate. The difficulties of ensuring that plausible assessments of particular forms of capital assets, particularly realty, are made by taxpayers are well known and the change in the system of income taxation is in part a 'finessing' operation. Although impacts on the valuation of forests are unclear at the time of writing, the illustrative calculations for a forest owner set out below assume the new system will not alter existing provisions in respect of forest value assessments.

Interaction between the Forest Trust Fund and income tax

The value of the tax provisions in respect of repayments from the Forest Trust Fund may be recorded at this point. An owner must pay between 5 and 25% of receipts from sales of wood, however sold, to the Fund. The current rate paid averages 12%. This is a deductible expense against tax. The money is collected by the owners' association in the case where sale is arranged by an association and this is then passed to the Forest Service. Unlike these parties, the owner gains no interest while money is with the Fund. The keen operator minimizes the interval between payment and claiming money back from the Fund for expenditure on allowable works, principally silviculture and roading. Others use the fund as a cushion, of use, for example, when grants are reduced. On average, however, money lay with the Fund for 1.5 years in 1989.

The crucial benefit of the scheme is that only a proportion of the repayment is taxable. This slice varies between 65% for repayments of 50,000 N.kr. or less and 95% for 500,000 N.kr. and above. Eid (1990) has estimated the percentage grant equivalent of this provision for different actual grant rates and marginal tax rates. For an operation with a 25% grant rate and with the current effective rate of income tax of 28% at the margin, the grant is enhanced to 35.2%, ignoring the lapse of time during which the money is out of the owner's bank account.

Income tax

Following the 1992 tax reform, the income from a forestry business owned by a self-employed person is split into capital income and personal income. Capital income consists of two parts: identifiable income from financial assets, such as dividends, and yield on business assets. As the summary of proposals made in 1991 explains (Ministry of Finance, 1992), the latter component is rarely expressed explicitly. This will be calculated administratively as the product of a capital yield rate (set initially at 16%) and a business capital yield base. A countrywide average for net annual cash income from the forest of 200 N.kr. ha^{-1} is assumed. Capital income is calculable at 16% of the capital value of the asset. If the forest value is set at 1000 N.kr. ha^{-1} the capital income becomes 160 N.kr. ha^{-1} and a marginal rate of 28% is chargeable on this. The average rate at a total capital income of 300,000 N.kr. works out at 26%. Secondly, the remainder of the net income, that is 200 −160 or 40 N.kr. ha^{-1}, is assessed to personal income at a higher rate. Assuming a marginal rate of 45.3% for the average forest owner implies an average of ca. 31% as the rate applicable to an income of 200,000 N.kr. The average income tax due thus amounts to $0.26 \times 160 \times 0.31 \times 40 = 54$ N.kr. $ha.^{-1}$ $year^{-1}$.

Property tax

Property tax is payable only on real estate in urban districts: its impact on private forestry holdings is negligible.

Wealth tax

This is payable at the rate of 1.0% for municipal tax and at rates varying between zero and 0.3% (over 825,000 N.kr. total wealth) for state tax. Formerly, the valuation of forest land was deliberately conservative, being based on a low annual yield capitalized at 10 year's purchase; this might have generated a typical value of the order of 1000–1500 N.kr. ha^{-1} (Eid, personal communication). Assuming that in practice the future valuation is 1,000 N.kr. (Bjorå, personal communication) and the average rate of tax applying to an average forest owner is 0.9%, the annual burden becomes 9 N.kr. ha^{-1}.

Inheritance tax

Maximum marginal rates of inheritance tax are 20% for family members and 30% for persons outside the family. Assuming an average rate of tax for transfer to a close relative at death certainly no higher than 10%, and again a capital value of 1000 N.kr. ha^{-1}, reduced in the case of assessment to this tax to 75% or 750 N.kr., the average charge amounts to 75 N.kr. ha^{-1}. The annual equivalent on a 25 year interval between transfers, adopting a multiplier of 0.03, is thus 0.03 × 75 or 2 N.kr. per ha^{-1} $year^{-1}$.

Summary of direct tax charges

The total tax payments on a countrywide average basis are thus (N.kr. ha^{-1} $year^{-1}$):

• Income tax	54
• Wealth tax	9
• Inheritance tax	2
• Total	65

GRANTS

The general principle of providing forest owners with public funds where there is otherwise likely to be a major divergence between social and private optima is probably applied with more rigour in Norway than most other countries. A great deal of *ex ante* economic analysis was undertaken in the 1960s, for example, when plans were being laid for the major increase in investment in forest roads especially. But there is clearly much room for political manoeuvring in the system of grants, many of which differ across counties and even municipalities.

Grants are usually stated in terms of percentages. (Details are set out in leaflets issued by the Forest Service; these are listed under 'Landbruksdepartementet' in References to this chapter.) The rates are maxima and grants are not demand-led as in Britain but allocated according to priority within a restricted budget. Provisions in recent years have been of the order of 350m N.kr. from two sources (Scheistrøen, personal communication). One is the normal forest department allocation, the other, operative from 1985, derives from the Agricultural Development Fund (Landbrukets Utbyggingsfond). Claims for grants are on the basis of approved schemes, many of which have to be approved by the nature conservation authorities. It may be noted in passing that several of the grants are available to the State Forest Service, i.e. the state forestry enterprise.

Plantation establishment

Highly specific terms apply for the various establishment operations, designed it appears to ensure that sensible and socially acceptable work is done. The maxima range from 25% to 60% of approved cost, depending on location, and the average is about 30%.

Inventory and management planning

Grants in this area account for a considerable proportion of the total spend. The inventory payments are made to favour surveys covering a number of owners' properties. The maxima are 64% for a set of not fewer than five owners owning only 10 ha and 44% for a single owner with 10 ha, both rates falling with increasing size.

Roads

About one-third of the total budget for grants is taken up by support of road building. Grants are conditional on a certain level of removals being carried out over an 8-year period.

1. Interest-free loans may be granted by the National Land Bank for between 50% (where single properties are affected) and 60% (where several ownerships benefit) of approved costs.
2. Grants of up to 40% in certain counties, rising to a maximum of 60% or even exceptionally to 75% are available: the average is about 30%.

Other operations

Grants for first thinnings and for the long-distance haulage of broadleaved wood are difficult to reconcile with a liberal economic argument for grant support. Grants are also available for extraction in difficult terrain. The rules for the latter are rather complex, rates from 1992 being set at maxima per hectare for different terrain classes, with maxima per cubic metre also applying. In the mid-1980s two-thirds of such grants were paid for work on terrain with a gradient of more than 40%, average grants per m^3 being 27 N.kr. on sites logged without a cable crane, rising to 63 N.kr. for sites where cable crane extraction was necessary (Inderberg, 1992). Regional employment and income considerations play a role in justification of such assistance, with forestry favoured because of a high labour content and, in part, the wish to curb expensive extra agricultural production.

Extension

Public service staff engaged on administration and extension in connection with state control of private forestry numbered 530 in 1991 less an element for other than private woodlands of, say, 10%. At an average annual cost of 300,000 N.kr. per person, this implies a cost of 143m N.kr. in 1991. In 1965, the proportion of county staff time on extension was 9.2% and that of district staff 11.2% (Director of Forestry's annual report, quoted in Svendsrud, 1969). In 1989 and 1990, extension accounted for 15% and 18% respectively of forest service input (Aalde, personal communication) including general administrative costs implying a current cost of extension work of the order of 25m N.kr. year^{-1}. Part of this spending applies to afforestation rather than to maintenance of the existing estate alone.

An important consideration is that work related to planning accounts for twice the effort devoted to extension proper. Although some of this effort may be seen as tantamount to extension work, part must be regarded as consultation. The breakdown of such work in 1990 was as follows: technical planning, 30%; management related to incentives, 24%; forest surveys, 21%; economic planning, 11%; public planning, 14%.

Research

Publicly funded research is carried out by the Norwegian Institute for Forest Research (NISK), the cost of which in 1990 amounted to 50.2m N.kr. Research conducted at the Norwegian Agricultural University's Forestry Department has been estimated to cost 7m N.kr. A particular emphasis of relevance to this study is the university department's strength in analytical work on both goals and means of forest policy. Assuming combined expenditure on research of 57.2m N.kr., it is considered appropriate to allocate 75% of this total to private individual owners, i.e. 42.9m N.kr.

Total Public Expenditure in Support of Private Forestry

Estimated public expenditure on private forestry in 1990–1991 is as follows (N.kr. year^{-1}):

- Grants 350m
- Extension 25m
- Research 43m
- Total 418m

Note: grant administration cost 66m and planning 52m. N.kr.

Outstanding Issues

Following publication of the Alstadheim report, a new statement of policy for Norwegian forestry may be expected. Several changes in the rules according to which forestry is practised have been made in the light of environmental requirements; no doubt more will occur. Among these, it appears quite probable that regulations will be made to control the logging of mountain forests, largely safeguarded recently by inaccessibility. At the same time the strict protection of arable and in-bye agricultural land is declining, 100,000 ha having been abandoned in recent decades, and it is generally thought appropriate that this land should be afforested. However, as already noted in relation to Ireland, farmers' organizations are concerned that the shift should not become iconoclastic.

The maintenance of rural activity remains both a concern and a problem. Total labour input in forestry, including delivery of wood to mills, only amounts to some 10,000 man-years[1], although total cut is over 10 m m^3. Afforestation of former (better quality) agricultural land will boost employment but the prospects are that employment must fall as labour productivity continues to rise. There remains the question of whether it is sensible, or even possible, to maintain the same distribution of forest ownership in the future.

It is widely held (see, for example, Alstadheim *et al.*, 1991; Wefald *et al.*, 1991) that society does better to support measures that underpin long-term supply than to subsidize current production. Norway led the way in recognizing the importance of roading in shifting the supply curve for roundwood rightward. Mention has already been made of a rationale for the subsidization of first thinning and logging in difficult terrain. Of greater importance in the long run is the continuing squeeze between costs of silviculture and declining standing tree values. Evidence on this point is found in, among others, Veidahl (1988). The problem is not discussed further here: its existence has already been recognized in relation to other countries.

Much attention continues to be paid to the raising of wood production, despite the warning of the recent price history. The level of cut continues to lie substantially below the allowable cut (or 'balansekvantum' as it is known, being the level of cut that can be assured without future decline). This difference averaged 20% over the last five years and the prospect is that a further 20% increase in allowable cut will be attained by the end of the century. These figures give cause for considerable concern over means of improving the mobilization of wood. Some grant aid is aimed at this goal, but another route lies in freeing up the market. In this regard it may be that the relations between the forest owners and industry with central price negotiation and conventions such as delivered prices for pulpwood being uniform regardless of location do not come as near to pure market competition as is practicable and desirable.

---

[1] From cooperative associations' compilations; national income statistics show a figure of 8000 man-years.

Objections to the highly mechanized forestry introduced during the 1970s mean that there is a growing emphasis in favour of smaller-scale technologies which are more labour-intensive. A shift towards such technologies holds no particular merit for forest owners, except where this matches their own tastes since contractors are often employed for the heavier work and such workers are likely to wish to operate the most cost-effective machines. The debate on this topic promises to be sharp.

There are two social considerations which may be noted. First, in certain areas of the country the feeling of the inhabitants and the political reality is nowadays that the maintenance of much infrastructure is no longer necessary or its cost acceptable. This applies in the far north of Norway where the islands have now become depopulated (McKay, personal communication). The same sort of pressure may also apply in less disadvantaged areas as in so many other countries. The process of adjustment will no doubt simply be somewhat slower in Norway owing to the strong political attachment to regional policy. Second, it is sometimes held that a diversity of decision makers in a community is desirable. This may be carried to extreme but one aspect of the argument, namely the restriction to landowners broadly to those of the particular commune as a result of the land laws, runs counter to this reasoning.

The Alstadheim committee noted the desirability of encouraging savings through the tax route. The desire to maintain a high savings ratio is familiar: Norway has long been known to display one of the highest ratios and also one of the largest capital–output ratios of any economy in the world. Whether this concern will continue to meet with general endorsement is unclear.

References

Aalde, O. *et al.* (1989) [Multiple use forestry: forestry's relation to the natural environment and outdoor recreation.] *Norges Offentlige Utredninger* Nou 1989, 10. Oslo.

Alstadheim, H. *et al.* (1991) [Norway's agricultural policy, demands, aims and means: main report.] *Norges Offentlige Utredninger* 1991, 2B. Oslo.

Anon. (1985) [Economic policy in forestry.] *Statens Meldinger.* nr.18 (1984–1985), Oslo.

Anon. (1988) *Act of 21 May 1965 Relating to Forestry and Forest Protection (the Forestry Act), as amended.* Eng. translation, typescript. Ministry of Agriculture, Oslo.

Anon. (1990) The Forest Trust Fund. Typescript. Forest Service.

Eid, J. (1990) The Forest Trust Fund. Typescript, Ås.

Eid, J. (1991) [What will tax reform mean for forestry?] Typescript, Ås.

FAO (1988) Forestry policies in Europe. *FAO Forestry Paper* 86, Rome.

Hamilton, H. (ed.) (Undated) *Nordic Woods*. The Swedish Forestry Association, Stockholm.

Inderberg, T. (1992) Public incentives to stimulate the cutting activity in difficult terrain in Norway. Typescript. Forest Service.

Landbruksdepartementet

(1984) [*Regulations for subsidy on first thinning*.]

(1984) [*Instructions on grants for road building*.]

(1985) [*Instructions on support of transport in forestry*.]

(1987) [*Instructions on grants for working plan preparation.]*

(1987) [*Instructions on investment loans for forest roads*.]

(1991) [*Instructions on state grants for plantation establishment and maintenance*.]

(1991) [*State grant for establishment and maintenance*.] Rundskriv M57/91.

Ministry of Finance (1992) *Norwegian Tax Reform: a summary of the main elements of the Norwegian Tax Reform Bill of 1992 presented to the National Assembly on 12 April 1991*. (Engl. translation.) Oslo.

Norwegian Forestry Society (1991) *Forestry in Norway*. Oslo.

Opheim, T. (1991) *Norwegian Forestry*. Paper to 25th Session of the European Forestry Commission, Oslo. Forest Service, Oslo.

Svendsrud, A. (1969) Some features of Norwegian forest policy. In: *Readings in Forest Economics*. Universitetsforlaget, Oslo.

Veidahl, A. (1988) [Some developments in Norwegian forestry.] [Forestry in the oil age.] *Rapport* nr.5B, Ås.

Wefald, O.J. *et al.* (1991) [*Annual Report 1990*.] Norges Skogeierforbund, Oslo.

World Bank (1990) *The World Bank Atlas 1990*. Washington, DC.

14

SWEDEN

BACKGROUND

The Swedish economy is an efficient one, producing a GDP per caput equal to that of Norway and the United States, and although its forests account for an area less than 1.0% of the world's area of closed forest, Sweden constitutes a major forestry power. Its output of industrial wood, though substantially lower than at its peak in the early 1970s, is the fifth highest in the World. Forest products exports amount to 20% by value of all exports and the country is the third largest exporter of paper and board in the world. Roundwood production potential is high: increment is some 91 mm^3 overbark, and the allowable cut, the more relevant figure for wood supply, is estimated at 77m m^3 during the 1990s, more than 10m m^3 higher than recent levels of cut. Allowable cut is increasing rapidly and according to the 1992 Forest Policy Committee (Lorentzon *et al.*, 1992) is expected to reach 95 mm^3 from the end of the 1990s onward.

The biological character of the country's forests is conveniently uniform for wood-using industry, but the aims of a diverse range of owners and the demands of others, especially for non-wood benefits from forest land, vary widely. Private non-company owners own half the forest land, the relative areas, in millions of hectares being; private 11.7, companies 5.5, state 4.5, other public 1.9; total 23.6 (Larsson, 1990). Most of the non-company holdings are in the south of the country and most of the state forests in the north. Because of this difference, the physical potentials of the various sectors differ and the allowable cut reflects this point. The dominance of wood production as the principal goal of forestry and the support of this function as the main occasion of government intervention in the past are clear. But these purposes are by no means regarded as exclusive: much attention has always been paid to the provision of non-market benefits and this concern continues to grow.

The distribution of forest ownership by size has its roots in history.

Whereas in Norway the peasants acquired Crown lands in the 17th century the evolution in Sweden was different. The position of the nobility as military commanders was reinforced by the successive wars under the leadership of Gustavus Adolphus and Oxenstierna to such an extent that the nobles gained two-thirds of the land and created manorial style estates (Tracy, 1989). In Denmark the peasants became serfs, but in Sweden, the legislative body, the Riksdag, contained the four Estates of nobles, clergy, burghers and peasants (Toyne, 1948). Some of the large landed properties remained with the aristocracy, some developed their forests and mines to evolve into the major industry combines of today.

The holdings of forests owned by wood-processing companies are very large. By the standards of manufacturing industry in developed economies generally, they are extremely highly concentrated with no less than 91% of their forest assets measured by area in the hands of the four largest: Svenska Cellulosa with 1.82m ha, STORA 1.58m ha, Mo and Domsjö 1.04m ha, and Korsnäs 0.52m ha. Because these are the major consumers of roundwood, this degree of concentration implies dominance of the demand side of the wood market and a very large influence on the supply side by single firms over large areas, both features that are particularly relevant to small owners in view of the market power the companies can exert.

The areas of sizes of the forest holdings of individuals are as follows:

Size-class	0–25	25–100	100–400	400 plus
Percentage of total area	13	40	36	11

Source: Figure B1, Larsson (1990). The total area on which the percentages are based is the 10.6m ha recorded by the agriculture and general property censuses in contrast to the figure of 11.7m ha from the National Forest Survey.

Farm forests are defined as forests worked in combination with farming conducted on an area of more than 2 ha. Though still an important element within the non-company sector, their extent has fallen dramatically from forming the bulk of the total in the 1930s and 1940s to less than 40% in 1988. Average sizes of both farm-forest and pure forestry holdings have increased slightly. Properties in parts of middle Sweden are often much fragmented, sometimes in quite narrow strips, and a survey in the 1960s concluded that 2m–3m ha would benefit from restructuring (Larsson, 1990). There were some 235,000 privately owned holdings in 1988, implying, because of multiple ownership, a total of some 300,000 owners or part-owners. More than half these owners do not live on their holdings but in urban areas (Swedish Institute, 1991).

Information is available from official sources and, in more detail, from specific sample and other studies on the structure of private holdings, owner characteristics, etc. A project team of the SIMS (Forest – Industry –

Market Studies) group of the University of Agricultural Sciences has produced assessments of the decision system of owners (Hansson *et al.*, 1990, Lönnstedt and Tornqvist, 1990), of changes in ownership structure of private woodlands (Eriksson, 1989, 1990) and of responses to different management strategies (Hansson *et al.*, 1988). These studies provide a valuable basis for judgements on the likely consequences of changes in the policy and general economic environments.

An important role in Swedish forestry is played by projections of the future possible cut of timber based on the application of alternative cutting regimes to stand data provided by the National Forest Inventory. This survey uses a system of continuous inventory with hidden plots and is carried out by the Department of Forest Survey of the Agricultural University at Umeå (von Segebaden, 1990). The naive observer might be forgiven for wondering why Sweden requires three surveys of forest land (there are also so-called economic land maps surveyed by the Swedish equivalent of the Ordnance Survey). The reason claimed for the national inventory is to have a less detailed but unbiassed inventory that covers all ownerships and allows consideration of the effects of different long-term strategies of thinning, felling, species choice, etc.

Forest land constitutes 65% of Sweden's land area (Sohlberg, 1990) and the value placed on it for recreation in the broadest sense is high. This is especially so in a country with the high income of Sweden, with low working hours, good transport facilities and a long history of economic dependence on and love of the forest.

A measure of the importance of the last point was provided in the 1980s when popular concern was extreme over the increasing weight of acid depositions and the damage caused to lakes and soils in Sweden. Anxiety continues to be expressed over the long-term effects of acid inputs on forest health. Measures are in hand to lime 8000 ha at a cost of 24m S.kr. over the years 1989–1990 to 1991–1992 (Wijk, 1991). The current careful approach to the subject of deleterious effects can be gauged by this reference and that of Skogsvårdsstyrelsen (Frederiksson, undated) which describes a number of initiatives for 'revitalising' forests in Blekinge County. The 1992 Policy Committee commissioned work by Warfvinge *et al.* (1992) on the impact of future pollution scenarios on critical loads of pollutants measured by the base cation to aluminium ratio and the expected effects on increment. The relationship between these two variables is based on 15 points for data derived from field experiments on increment of *Pinus* and *Picea* spp. across the world. For plausible pollution emission scenarios, the experts' prognoses suggest that total increment could be reduced by about 10%. However this prognosis must be viewed with caution; the assessment represents the first comprehensive attempt at estimating this all-important figure and the recency of the analysis means that it has not been subject to critical scientific appraisal.

Policy and Legislation

Evidence of the national love of the forest was strongly evinced in the 1980s when commentators drew attention to the threat posed by air pollutants to the health of the forest. Combined with the attitude of Swedes of concern for the social fabric, the love of the forest combines to establish a clear line of policy towards forestry. In the 1970s forestry's purpose was obvious. Thus the aim of forest policy, still current in 1992, was clearly stated in Section 1 of the Forestry Act 1979 (Skogsstyrelsen, 1991a):

> By means of proper utilization of their wood-producing capacity, forest land and forests growing on them should be managed in such a way as to provide a permanently high and valuable timber yield. This management should pay heed to nature conservation and other public interests.

The Swedish Institute (1991) develops the theme:

> The forestry sector is a collective term for forestry and the forest industry. . . . Underlying Swedish forest policy is the conviction that there will continue to be a demand for forest products in the future and that the forestry sector will continue to be of major importance for the country. One of the implications of this is that forest owners are legally obliged to apply a certain minimum standard of silviculture.

None of this is unfamiliar in European forestry: what is remarkable is the degree of control implied and growers' acceptance of this.

Since the 1979 Act, there have nevertheless been a number of developments, often leading to new laws, which significantly qualify the productivist theme of the 1979 Act. The Broadleaved Deciduous Forestry Act (Skogsstyrelsen, undated) was one such, which coincidentally came into force in the same year, 1984, as the announcement of Britain's broadleaved policy. It qualifies the rather general objects in the second sentence of the 1979 Act thus: 'The aim of this Act is to preserve the nation's deciduous forests for posterity.'

An earlier but highly relevant law passed in 1964 was the Nature Conservancy Act which sets out stricter provisions for sites of special conservation interest. As in Britain, the official nature conservation agency has never been backward in promoting its interests: some 11% of Sweden's land surface was covered by one form or another of nature conservation designation in 1989 (Dahlin, personal communication).

The Forestry Act 1979 resulted from the work of a government commission appointed in 1973 to study the forestry sector; after a period of extensive consultation this led to preparation and passage of the bill. The position then was that policy was dominated by the desire to provide increasing quantities of good quality wood to industry. Since that time the market demand for wood has risen only slowly, running at about 66m m^3 over bark standing during the 1980s, while increment and allowable cut have increased almost continuously. Environmental demands on the forest have meanwhile risen until they are viewed in official quarters as being 'on a level with wood production requirements and no longer as a constraint on the cost-effective production of raw material'.

Before turning to the most recent expression of environmentalist attitudes to wood production, it is relevant to note three further features of the legal and policy framework which influence foresters' actions. First, as in Norway, there are important laws restricting the freedom of those wishing to purchase land. Outside the thinly populated development areas, a private individual does not require a permit to buy land. For others such as forest products companies the rules are very restrictive. The reason has its origins in fears that firms would exert their market power by depending more heavily on wood from their own holdings when prices were high thus disadvantaging small forest owners. In the development areas, a permit to buy is not required if the purchaser already lives in the same municipality or undertakes to settle on the property for at least 5 years. Otherwise permits are refused to any intending purchaser and arrangements may then be made for the state to acquire if the vendor demands.

The control of landownership change has its own special emotional as well as economic aspects. The second general point is that an attitude in favour of control rather than the reverse extends to almost all other subjects influencing forest management. Thus the various laws bearing on forestry are supplemented by a substantial array of regulations and instructions. This feature is closely allied to the support given in Sweden to a generally interventionist philosophy which was for a long time the central plank in the platform of the Social Democrats. 'Freedom with responsibility' as a rule of conduct applies in Sweden as well as in Norway, but it appears that the responsibility in the former is rather more burdensome. The Swedish willingness to accept order in support of the general weal is different from the love of order found in some other countries. It arises only after much debate and thought reflecting concern and responsibility for the whole of the community. In some fields such as silvicultural practice, public discussion is naturally less widespread. In these circumstances, what the 'authorities' say tends to be accepted as politically correct (in the 1990s sense of that term). As of 1987 it was possible for a commentator (Johannisson, 1987) to describe the powerful institutions as 'creating' a corporatist state - hardly a new conclusion - although at the same time locally based initiatives were establishing many new businesses. He noted that the ideological climate was more favourable to smaller scale industry in the 1980s but that capital taxation and the demand for social security funds gave cause for concern. Hultkrantz (1991), in a strong critique of the philosophy underlying forest policy, advocates a departure from the State's post-war obsession with greater wood production and the regulatory framework created in its pursuit. Instead, he argues, the need is simply that the state should concentrate on the protection of the general social interest in forests. It is likely that recent changes in national politics will herald a new assessment of the need for the attitude that may be epitomized as 'Nanny knows best'.

The third issue influencing forest policy is that of energy policy. The phasing out of nuclear power generation poses major problems. Conservation and landscape considerations limit the potential of hydro-power as a substitute, and fossil fuel-powered stations are not favoured. In these circumstances the possible contribution of wood sources of energy, both from forest and other waste as well as plantations, becomes interesting. Hultkrantz (1989) discusses this with special reference to pulping and the impact of energy price on this industry. An alternative less likely possibility is to reverse the decision on nuclear power and to use the forest as a means of locking up ever more carbon as two major contributions to slowing the accumulation of atmospheric carbon (Hultkrantz, personal communication).

Policy Review of 1992

A parliamentary committee was appointed in 1990 to evaluate existing policy and to make new proposals for policy. The main problem facing the committee was that of identifying the appropriate balance between production of wood and environmental benefits. Global issues, biological diversity and the risks of global warming were part of its remit. In line with the philosophy of the government, attention was to be given to the possibility of more self-financing of activities required by society. The review may be expected to lead to new legislation in due course; it is clear however, that the committee's views are subject to strong criticism from several sides.

The committee had the impossible task of reconciling the two interests of maintaining and improving the environment on the one hand and earnings on the other, matters which affect Sweden perhaps more strongly than any other of the European countries reviewed here. Traditionally, forests formed a well-balanced whole, providing workers and owners in both forest and wood-using industry with jobs and earnings while maintaining a forest structure which was perfectly acceptable in terms of impact on nature conservation and landscape. The past 30 years have witnessed possibly the most determined and sustained campaign to increase efficiency in forestry that may be observed in any European country. In place of the traditional, gentler style of silviculture and harvesting, modern large-scale fellings using two-shift logging operations, mechanical site preparation, fertilization, draining, herbicides and more exotics (notably *Pinus contorta*) have combined to generate a powerful backlash.

The Swedish nation is a people whose roots lie in a richly forested countryside: their feeling of unease over forestry practices that actually mean something to the man and woman in the street is a matter which politicians ignore at their peril. The concept that forests are quintessentially *natural* resources automatically loads the debate about their proper management with special meaning. Against these considerations, costs and prices have been kept low, so allowing consumers to enjoy the benefits of efficient forest and forest products industries, national product has increased and the economy has gained in stability through a widening of its manufacturing base, income has been generated in disadvantaged areas such as inland Norrland where jobs could otherwise only be provided at high public cost, and a major contribution has been made to the overseas balance.

The relatively static level of cut in latter years reflects a slow growth of demand for softwood products worldwide and hence static or declining prices, an unwillingness to cut on the part of some and an inability of others to supply wood more cheaply. Noting this failure to harvest the allowable cut, certain observers recognize in the current position an opportunity to claim that the intensity of forestry practice remains too high and may be reduced, and in the process produce more environmentally attractive forests.

However plausible this line of argument appears it contains a major difficulty, even fallacy, which underlies the omnipresent debate about the balance between market-driven activity and the environmental movement's utopia, a large area of the countryside given over to conservation. The difficulty is the greater where the commercial contribution is as large as it is in Sweden and accordingly the trade-off between reduced cut and increased conservation is more difficult to decide than in, say, Flanders or southern Britain. The danger is that a view considering the situation over the succeeding 5 or 10 years, or a fraction of a typical Swedish rotation, carries no implication that the relative values of commercial and non-market outputs of the forest will not alter in another period towards wood production. Forestry always stands in this curious position of having to face the problems of the short term while the resource that is at issue is measured in terms of its value by the revenue it produces over the long term.

The Forest Policy Committee's report takes a somewhat ambiguous path between on the one hand the relaxation of checks on forest practices aimed at the production of wood and, on the other, the idea that measures to increase conservation values are necessary and may justifiably be encouraged by the state. It proposes that the 1979 law should be dramatically amended in respect of the balance. Although, as noted below, the report offers a number of pacificatory gestures to environmentalists and more conservative foresters, it balances these with the neo-liberal concept that adjustments on private land to meet nature conservation goals should be achieved through voluntary action.

Administration

Sweden's national regulatory body, the National Board of Forestry (Skogsstyrelsen), is part of the Ministry of Agriculture but it operates with a high degree of independence (Skogsstyrelsen, 1991b). It directs and co-ordinates the work of the 24 County Forestry Boards. Together the two levels of the organization, central and regional, are known as the Forestry Organization (Skogsvårdsorganisationen, SVO). Since 1981, the organization has been conducting a General Forest Inventory (Översiktigsskogsinventering or ÖSI) which covers all private forests and will be completed in 1994. It is mainly aimed at short-term, operational planning, being used by County Boards in the administration of forestry regulations. It can thus assist in the identification of the effectiveness of various policy measures (Swedish Institute, 1991). The recent forest policy committee doubts whether the benefits of the information compiled are matched by the survey's costs.

In contrast to the position in other countries reviewed in this study, the state forest service (Domän, previously Domänverket), is part of the Ministry of Industry, an interesting reflection of the weight given to commercial wood production in Sweden. In January 1992 the enterprise became a corporation with its shares held, as in the case of Ireland's Coillte Teorante, by the relevant Minister. The government subsequently indicated its intention to sell the company.

The main task of the National Forestry Board has been summarized as being 'to ensure that the condition and yield of Sweden's forests satisfy national forestry policy' (Ekholm, 1991). Again the emphasis on (wood) yield is notable. The functions performed (Skogsstyrelsen, 1979) are to advise government on policy development and legislation, to supervise owners, including the State, in order to ensure that forest management is undertaken as required by the forestry laws, to offer advice and information to owners, to administer grants, to conduct or support necessary enquiries, to provide courses for foresters and the general public, to assist other public authorities concerned with relief work, nature conservation jobs, etc, and to provide certain contracting and consulting services. The last include plant production, marking timber and the preparation of management plans, activities which have grown to a size which makes SVO a considerable trading organization in its own right. In forest management planning, SVO has a market share of 60% of the country total (Skogsstyrelsen, 1991b). Total expenditure is almost exactly equally financed by national grants, commissions from other public bodies and earnings from sales of plants and services.

The dominating role of Skogsstyrelsen including the command it has through the commercial activities it manages has been subjected to increasing criticism in recent years. Thus Hultkrantz (1991) identifies a number of areas where the degree of regulation is not only unnecessary but actually causing losses to growers and to society. An example of evaluation of the economic impact of control, in this case the strict cutting rules applied to forests, is provided by Lönnstedt (1991). He found that the nationally recommended management strategy theoretically results in a total (capitalized) loss of 100–150 billion kronor. It is hardly unexpected to find the Forest Policy Committee condemning a number of interferences with owners' choices of action; it is disturbing at the same time to find that the Committee favours new but arbitrary guidelines such as the exclusion of 5% of the total forest area from forest management in favour of its designation as nature reserves or national parks. Proposals to reduce the areas devoted to *Pinus contorta*, to restrict the use of fertilizer severely, to favour the use of broadleaved species on setaside land, etc. are not surprising given the starting point of the review. What is disturbing is the apparently arbitrary choice of goal and the various technical means of achieving it.

The national nature conservation agency (Statens Naturvårdsverk) is currently obliged to leave to the Forestry Board responsibility for the administration of nature conservation matters on forest land, apart from specially designated areas. Its cool reception, as recorded by Lindevall (1992), to the policy committee's proposal that conservation be left to the voluntary action of forest owners raises the prospect of strenuous inter-agency in-fighting.

Since 1983, owners of over 20 ha have been required by law to have a full management plan, smaller units being allowed to use a simplified form of plan. However, Sennblad (1988), who surveyed 3000 owners in 1984, estimated that only 32% of owners, holding 51% of the private forest, in fact had a management plan and nearly 30% of these were more than 10 years old. It is understood that the percentage of area covered had risen to 60% by 1991. As part of its trading-off of owners' interests against the imposition of tasks which are not subsidized or otherwise compensated, the 1992 Policy Committee recommends that owners should no longer be required to produce a working plan, which is regarded as being an owner's responsibility. Instead owners should be statutorily obliged to provide authoritative accounts. These, it is suggested, are necessary for the execution of policy and for social planning. Sampling is not proposed to fulfil what is otherwise an interesting departure from the conventional physical basis of so much forest planning.

Private forest owners have very effective producer cooperatives of owners who collaborate through a national federation of forest owners (Skogsägarnas Riksförbund). In 1990, total membership of the cooperatives amounted to 86,446 holding 5,488,000 ha or roughly half the area of all non-industrial private forest. This area and the cut from it both represent approximately a quarter of the country totals. The usual functions of owners' organizations are performed; namely price negotiations, representation of members' interests in policy discussions and negotiations with government and other bodies, roundwood trading, contracting services, information and training for members. The function of agreeing prices with forest products industry remains the dominant concern of the associations. Here again the 1992 Policy Committee has an important observation to make; in order to ensure that the market functions smoothly it is considered that the market should not be concentrated through either a buying or selling organization in the various regions of the country. Currently, as a definite policy of aiming to influence prices, some of the Associations have mills of their own. In recent years, their output accounted for 10% of Sweden's total for forest products (O'Laughlin and Messina, 1988). Overall, the associations' organization on the forest side is large, with 294 districts over the whole country at the end of 1991. The average size of forest holding per member is close to that of the whole country. A pragmatic attitude is taken on the issue of the needs of small owners; economic pressures inexorably demand that the services of cooperatives be competitive with those of large contracting firms and some district mergers are likely (Österblom, personal communication). There is no obligation for members to sell to the mill of their own or any other association: this has led at times to embarrassing roundwood supply problems.

Implementation

Discussing public policy toward private forestland in Nordic countries, Marsh (1954) summarized the position thus: 'Sweden has a closely integrated national program for private forest land, embodying regulation, individual service, general education and extension, grants-in-aid and forest credit'. The 1948 Act in operation when Marsh made this comment was based on the concept that the forest owner should be enabled to achieve an economic return from his forest. Although the Act heralded a general tightening of control there was no compulsion and it was only with the passing of the 1979 Act that certain minimum levels of practice were established. Sweden has continued to deploy the full array of publicly financed aids combined with the many regulations to which reference has already been made. As already noted, a number of published critiques have questioned the economic merits of certain of the regulatory and financial interventions. The policy committee adds its view that several of the latter have been ineffective or had undesirable side effects. It therefore welcomes the fact that some have been progressively abolished in recent years, especially as the desire to encourage more investment in silviculture for wood production has diminished, the anxieties of nature conservationists have been noted and 'liberal' political ideas have come to have an increasing influence.

The 1992 Policy Committee made no contribution on taxation matters, nor was it asked to.

TAX

Taxes and social security contributions in Sweden accounted for 66% of GNP at factor cost in 1989 (57% of GDP at market prices), the highest rates in the world (Central Statistical Office, 1992). The comparable figures for the United Kingdom were 43% and 37% respectively. Similarly direct taxes on households and social security contributions in Sweden account for the highest proportions of total personal income of any country in the world. Against this background, the authorities are posed an interesting test of such principles as equity and neutrality in assessing forest owners to tax. Further alterations to the system introduced in 1979 occurred following the installation of the new government in 1991. So far as income from a business is concerned, the new arrangements, which include a broadening of the tax base and a reduction of rates, imply in general a reduction in income tax. (Ministry of Finance, 1991 forecasts a loss of 89 billion S.kr. of total revenue from income tax through rate reduction, but increases in revenue of 13 billion S.kr. from employment income, 39 billion S.kr. from capital income and 28 billion S.kr. from VAT.) Because of the novelty of the changes, it has not proved practicable to carry through calculations on the new basis, and the system of taxation described here is that obtaining up to 1991 (Skogsstyrelsen 1989, 1991c).

A fundamental role in taxation of woodlands is played by the measure of 'forest value' (skogsbruksvärdet). Real estate value assessment is undertaken on a rolling basis, formerly every five, now every six, years. The latest round began in 1988 and forest together with farmland was to be valued in 1992. In 1981 the total value ascribed to all forest land was some 54 billion S.kr. or 2300 S.kr. ha^{-1}. Values are likely to rise to 6000 S.kr. on the 1992 valuation (Holm, personal communication). The value adopted is 75% of the ascertained market value, the reduction being introduced in order to cover any deficiencies in the valuation method as applied in practice. In addition, the government decided in 1981 to reduce the value still further to 90% of the result. It is believed that actual market values in the south of the country may be several times the tax value.

Income tax

The underlying principle of taxation of the forest up to 1979 was that the yield alone should be taxed. If an owner could prove that the forest's value had fallen during his occupancy, he was entitled to deductions against his income, but this led to difficulties in practice. From 1979, a simplified basis was introduced allowing deductions only against the income of the particular year up to specified limits. These deductions were then added to any capital gain over the owner's period of ownership and rendered liable to capital gains tax on transfer by sale.

In addition to this system of forest deductions, special provisions applied in recognition of the facts that forest revenues may be 'lumpy', that expenses for logging may arise in a different year from receipts, that expenses of restocking are heavy and arise several years after receipts from logging, and that wood prices fluctuate markedly. A further concession was that fuelwood used on the holding was not subject to income tax. Costs of silvicultural, roading and other operations and interest on loans financing such operations and the purchase of woodland were deductible on various bases of write-off.

The Forest Owners' Associations no longer compile accounts for a sample of members, but it is believed appropriate to assess average annual net profits at a figure of 200 S.kr. ha^{-1} of all private forest. It should be borne in mind that the average 1991 rate, including social security charges, on an income of more than 190,000 S.kr. was 63%. For working purposes it is assumed that the net effect of the various allowances and deductions is equivalent to the charge of an effective tax rate of 35% on the average forest enterprise income before deductions, i.e. 70 S.kr. ha^{-1} $year^{-1}$.

Property tax

Property tax was not charged on forestry buildings.

Forest fee (skogsvårdsavgift)

This charge, which was abolished as from July 1992, stands to be included in this assessment. It was calculated at 0.8% of the forest value, and was also treated as a cost element in the enterprise profit. Assuming a forest value of 2300 S.kr., its cost was then $0.008 \times 2300 = 18.4$ S.kr. ha^{-1}.

Wealth tax

Forest property values were reduced by 70% for wealth tax purposes. Thus on average the value became 0.3×2300 or 690 S.kr. ha^{-1}. The wealth tax regime appears to be onerous, rising in theory to a maximum marginal rate of 3.0% but its total impact was limited to 75% of the total of enterprise and capital income combined. In theory, the provision of a special form of valuation, adjusted in various more or less generous ways, would be expected to cause some shifting of the tax and hence appreciation of the category of property concerned. However, owing to the difficulties of purchase of forest land by outsiders, it is probable that the incidence of such switches, as well as the use of the various means of financing that can be adopted in order to reduce net wealth, was small. In practice, on an average-sized woodland of 100 ha, with outstanding loans and splitting between spouses, the tax burden is zero.

Other capital taxes

Capital gains tax

The Forest Owners' National Federation estimates (Hogfors, personal communication) that about half the transfers of forest property are made by gift or bequest, about a quarter by sale to a family member and the remainder by sale to a third party. If a sale is made, capital gains tax may be incurred. The gain is not indexed. Because of the burden of this unindexed tax, some provision has thus to be made for capital gains. On a market value of say 8000 S.kr. the gain over 25 years, say, with market prices inflating at 5% is 8000 $(1 - 1/1.05^{25})$ or 5638 S.kr. Tax at 30% implies a charge of 1691 S.kr., which annualized equals 52.8 S.kr. For working purposes it is assumed that 50% of transfers are by purchase/sale giving an addition for CGT of 26.4 S.kr. ha^{-1} $year^{-1}$.

Inheritance tax

The figure calculated here is for tax on bequest. The valuation of forest for inheritance tax purposes was also reduced by 70%. The charge is assumed to arise every 25 years for transfers from parent to child. Allowance has to be made for postponements that are allowed by the rules; it is assumed that these implied a delay effectively diminishing the present value of the charge to 90%. A tax rate on the average of transfers received from an owner of 25% is assumed. Adopting a multiplier of 0.0312 to yield an annual equivalent, the tax is $0.25 \times 0.3 \times 2300 \times 0.9 \times 0.0312 = 4.8$ S.kr. ha^{-1} $year^{-1}$.

Total

It is arguable that part of the value of the estate of the dead person can be ascribed to the profit from selling his forest holding and thus a charge to inheritance tax arises as well as CGT. Since the inheritance tax charge is small this is ignored here and a mean of 15.6 S.kr. ha^{-1} $year^{-1}$ is assumed.

Summary of direct tax charges

The total of direct tax burdens is thus estimated as follows (in S.kr. ha^{-1} $year^{-1}$):

• Forest fee	18.4
• Income tax (after allowable expenses, etc.)	70
• Property tax	0
• Wealth tax	0
• Capital gains tax/inheritance tax	15.6
• Total	104

Grants

Grants are paid at fixed rates or up to specified maximum percentage levels and at different rates for different regions. Those applying in 1991/92 (Dahlin, personal communication) are set out below.

1. Restocking in forest development area:	
(a) normal restocking – inner zone only	0.5–1.3 S.kr. per plant
(b) supplementary restocking operations	1.0–1.6 S.kr. per plant
(c) beating up after heavy rabbit damage	2.3–2.9 S.kr. per plant
rest of Sweden	0 S.kr. per plant
2. Broadleaved management in whole country:	
(a) broadleaved planting: all areas	2 S.kr. per plant
(b) fencing broadleaved planting	800 or 1500 S.kr. per ha
(c) supplementary BL regeneration operations	80%
(d) early thinning of broadleaves (oak, beech)	60%
(e) management plan for broadleaves	100%
(f) marking of trees, advice	free of charge
3. Approved environmental and cultural operations:	70%
4. Insect protection required by law:	grant for each specific case
5. Miscellaneous operations when grant paid per plant:	
(a) weeding	0.1 S.kr. per plant
(b) draining	0.1–0.2 S.kr. per plant
6. Road building:	loan guarantee to 40–75%
7. Other operations:	
(a) establishment of seed orchards	100%
(b) harvesting, storage and protection of wood damaged by storm, etc.	loan guarantee
8. Afforestation of farmland; no grant available for planting.	

Payments both in total and under the various heads have varied very greatly over the past decade. A major reason for the overall fall from 340m S.kr. in 1981–1982 and the peak sum of 431m S.kr. in 1982–1983 is that grants are no longer paid for rehabilitation of the 'less productive' stands. This recognizes their nature conservation interest. Secondly, more is paid under the Broadleaved Act and a similar amount in aid of nature conservation measures. Thirdly, grants continue to be paid on forestry operations in the development area or the inner part of it. Total grants were in 1989–1990: 273.7m S.kr.; 1990–1991: 264.4m S.kr. Provision for 1992–1993 was 89.8m S.kr. made up as follows: regeneration operations in inner Norrland 58m S.kr., broadleaved management 15.8m S.kr., conservation operations

16m S.kr. It should be noted that the above figures relate to payments to all forest owners for regular forestry operations: additional sums are expended by Skogsstyrelsen for activities such as inventories of various kinds (key forest habitats, swamp forest), work on the 'Richer Forest' project, and energy forest plantations.

The Policy Committee proposes substantial adjustments in the distribution of state grants. These would centre on planting of the so-called noble broadleaves (roughly all hardwoods except birch, alder and aspen), liming to counter soil acidification, planning of road networks, protection of particular habitats and a doubling of the amount set aside for acquisition of reserves.

Extension

Extension work carried out by Skogsstyrelsen has four target groups; first forest managers, second owners, third workers, fourth the public. Not all of this is provided free. However, out of a total expenditure of the order of 1000m S.kr., the cost of extension services was 50m S.kr. in 1989–1990 and 53m S.kr. in 1990-1991.

Research

Bäckström (1990) states that 300m S.kr. were available for forestry research in 1989–1990. The proportion of all private forest area to the country's total is 73%, and for both growing stock and cut the percentages are higher. It is assumed that the part of total research expenditure allocable to private forest as a whole is 70%.

Summary of Public Expenditure in Support of Private Forestry

Total public expenditure in 1990–1991 ascribable to all private forest is estimated to have been as follows (m S.kr.). It should be noted that the intensity of extension work and research attributable to individuals' ownerships is probably higher than for firms.

- Grants 274 (and falling)
- Extension 50
- Research 210
- Total 534

OUTSTANDING ISSUES OF POLICY AND IMPLEMENTATION

Past implementation measures to serve the aims of the 1979 Forestry Act must be regarded as generally successful in achieving the wood growing and production objectives of that law. In the early 1990s, the main issue affecting private forestry is the extent of adjustment intended to achieve the government's goal of 'richer' forest, meaning forests that are at the same time more diverse and producing more environmental benefits. Owners have received less support in recent years, their growing stock has increased, though not so fast as if previous policies of intensification of wood production had been continued, and their incomes from sales of timber have declined. The situation is that there has been a change of national policy *de jure* as reflected in, for example, the 1984 broadleaved forestry law, and *de facto* by progressive changes in grant payments. The state has moved from favouring a profitable operation in money terms, based on encouragement of the maximization of high-value wood production, the core of the 1979 Act, to a position with no clear goals and relatively little cash supplied either to compensate owners or, where appropriate, to stimulate the production of more non-market benefits. The 1992 Policy Committee recommends some change in this position, stressing the need to continue with this shift. It proposes a blend of relaxation of certain constraints on owners' freedom of action with increased restrictions on such operations as large-scale felling, drainage and fertilization, plus new or increased incentives. While the package proposed may be a moderately well-balanced one, and politically acceptable in the round, it should be recognized that the mix presented constitutes a highly pragmatic solution. The committee's solution appears to owe much to political horse trading: the report contains no objective assessment of the costs and benefits to the Swedish nation of alternative balances. This situation cannot be conducive to the generation of lasting confidence among woodland owners.

It has been emphasized that there has been an increasing divergence between actual and allowable cut. Associated with this and the low prices of wood of recent decades, there has been a growing tendency for owners to live off-farm. Lönnstedt and Roos (1990) record the fact that 85% of absentee proprietors report that their forest income is less than 10% of their net income (the corresponding figure for farmers is 50%). This raises the question of such owners' motives for holding forest and their attitude to the achievement of wood-production goals. On balance, this shift appears likely to be relatively favourable to the changing balance of forest management for conservation relative to wood.

A lesser issue but of widespread importance centres on the reconciliation of ever-increasing demands by hunters with provision of the nature conservation values that underlie the demand for a 'richer forest'. Although Mattson (1989) has shown how economic aspects of hunting and control of the stock of game can be handled, the absence of evaluations of conservation benefits is a drawback to establishing more objective assessments of appropriate courses of action.

Changing attitudes stemming from major changes in public and political opinion concerning the role of the state will come to have increasing relevance to the choice of policy tools as well as aims. A topic which may come to have some bearing on the competitive position of owners is the proposed privatization of Domän. Also the retention of certain commercial activities by Skosgsstyrelsen will need to be considered.

It seems likely that more decisions will be questioned concerning the rules that apply to individual properties, as well as those, such as the management of the growing stock, which apply to large numbers of owners in a region. In this connection, the Policy Committee's insistence on continued efforts to inform and train by means of extension programmes is realistic.

References

Bäckström, P.O. (1990) [Swedish forest research in the 1990s.] *Sveriges Skogsvårds Tidskrift* 3, 12–21.

Central Statistical Office (1992) International comparisons of taxes and social security contributions. *Economic Trends* 459, 115–126.

Ekholm, I. (1991) Swedish forestry as an example of national organizational structures and activities. Paper to FAO/ECE/ILO Joint Committee Workshop, Budapest. Typescript.

Eriksson, M. (1989) [The change of ownership structure in private forestry in a historical perspective.] *SLU/SIMS Report* no. 5. Uppsala.

Eriksson, M. (1990) [The change of ownership structure in private forestry prospects for the future.] *SLU/SIMS Report* no. 3. Uppsala.

Frederiksson, G. (Undated) [*Forest revitalisation*.] Skogsvårdsstyrelsen i Blekinge, Ronneby.

Hansson, K., Lönnstedt, L. and Svensson, J. (1988) [Tax situation for private Swedish forest owners.] *SLU/SIMS Report* no. 2. Uppsala.

Hansson, K., Lönnstedt. L. and Svensson, J. (1990) [Decision support system for nonindustrial forest owners.] *SLU/SIMS Report* no. 12. Uppsala.

Hultkrantz, L. (1989) The consequences for the forest sector of nuclear power discontinuation in Sweden. *Scandinavian Journal of Forestry Research* 4, 395–406.

Hultkrantz, L. (1991) [Aims and methods in forest policy.] *Ekonomisk Debatt* 3, 282–7.

Johannisson, B. (1987) Entrepreneurship in a corporatist state: the case of Sweden. In: Goffe, R. and Sean, R. (eds), *Entrepreneurship in Europe – the Social Process*. Croom Helm, Beckenham, 131–143.

Larsson, U. (1990) Who owns the forests? In: Nilsson, N.-E. (ed.), *The Forests*. National Atlas of Sweden, SNA Publishing, Stockholm, pp. 118–121.

Lindevall, B. (1992) [Nature protection agency: '(It is) obvious that environmental aims cannot be secured by the voluntary route'.] *Skogen* 8, 7.

Lorentzon, S.E. *et al.* (1992) [Forest policy before the year 2000.] SOU 1992: 76. Stockholm.

Lönnstedt, L. (1991) *Economic Consequences of the Swedish Forest Policy.* Department of Forest Industry Market Studies, Uppsala. Typescript.

Lönnstedt, L. and Roos, A. (1990) [The economy of the absentee non-farmers.] *SLU/SIMS Report* No. 15. Uppsala.

Lönnstedt, L. and Törnqvist, T. (1990) [The owner, the estate and the external influences.] *SLU/SIMS Report* No. 14. Uppsala.

Marsh, R.E. (1954) *Public Policy Toward Private Forest Land in Sweden, Norway and Finland.* The Charles Lathrop Pack Forestry Foundation, Washington, D.C.

Mattson, L. (1989) Hunting in Sweden. Swedish University of Agricultural Sciences, *Department of Forest Economics Arbetsrapport* 95, Umeå.

Ministry of Finance (1991) *The Swedish Tax Reform of 1991.* Ministry of Finance, Stockholm.

O'Laughlin, J. and Messina, M.G. (1988) Swedish forestry and forest policy. *Journal of Forestry* 86(7), 17–22.

Sennblad, G. (1988) [Survey of logging and silviculture in non-industrial private forestry in Sweden 1984; Part 2. Private forest owners and their holdings in Sweden in 1984]. *SLU: Institutionen för Skogsteknik*, Rapport nr. 176.

Skogsstyrelsen (1979) Sweden's new forest policy. Typescript. Jönköping.

Skogsstyrelsen (1989) Short report on Swedish taxation rules and their relevance for forest policy. Typescript. Jönköping.

Skogsstyrelsen (1991a) *The Forestry Act 1979.* Jönköping.

Skogsstyrelsen (1991b) The Swedish Forestry Administration. Typescript. Jönköping.

Skogsstyrelsen (1991c) [*Forest and tax.*] Jönköping.

Skogsstyrelsen (undated) *The Broadleaved Deciduous Forestry Act.* Jönköping.

Sohlberg, S. (1990) Forest mapping. In: Nilsson, N.-E. (ed.), *The Forests.* National Atlas of Sweden, SNA Publishing, Stockholm, pp. 131–137.

Swedish Institute (1991) *Forestry and the Forest Industry.* Fact Sheet on Sweden, Stockholm.

Toyne, S.M. (1948) *The Scandinavians in History.* E. Arnold, London.

Tracy, M. (1989) *Government and Agriculture in Western Europe 1880–1988.* Harvester-Wheatsheaf, Hemel Hempstead.

von Segebaden, G. (1990) The national forest inventory. In: Nilsson, N.-E. (ed.), *The Forests.* National Atlas of Sweden, SNA Publishing, Stockholm, pp. 132–133.

Warfvinge, P., Sverdrup, H., Ågren, G. and Rosén, K. (1992) [Effects of air pollution on future increment.] Appendix 16 in: [*Forest policy to the year 2000.*] (Report of forest policy committee.) SOU 1992: 76. Stockholm.

Wijk, S. (1991) Liming of forest soils in Sweden. Skogsstyrelsen, Jönköping. Typescript.

15

PRIVATE FORESTRY CONSIDERATIONS IN OTHER SELECTED COUNTRIES

INTRODUCTION

The study of almost every country reveals distinctive features either of its policy towards private forestry or the administration of that policy. Because of their common cultural heritage with Britain and their attitude to marrying science with utility, including the wide use of plantation techniques, Australia and New Zealand are of obvious interest. The discussion concerning these two countries is not intended to describe their policy aims and methods of implementation in any detail but instead to identify particular features of private forestry in each. In the case of eastern Europe and Russia, the recent and continuing political changes carry major implications for the organization of forestry, including the ownership of forests. The picture is in a state of rapid evolution, but because of proximity and in certain casės the likely closer linking through membership of the European Community, it has been considered useful to note the position in these countries as of 1992.

I. AUSTRALIA

BACKGROUND AND ADMINISTRATION

Indigenous forests cover 41m ha, 5.3% of the land area, of which 36.3m ha are regarded as productive. Some 11m ha are private. In addition 'woodlands', implying open lightly tree-covered terrain, account for 64m ha. There is widespread concern over the extent of forest destruction in the continent and the degradation of both remaining scattered growth and soils. Although in 1990 plantations accounted for only 905,000 ha, or 2.5% of the productive area, they supplied one-third of all logs. Of the plantations 30% are private, most being owned by wood-processing industry.

Under the Australian constitution, State and Territory governments are primarily responsible for the management of lands. The federal government has, however, played an increasingly important role in recent years. Thus it has been instrumental in establishing a Resource Assessment Commission, the nomination of the north east Queensland tropical rain forest for World Heritage status and, through the issue of export licences, control of the export of unprocessed wood. The Australian Forestry Council was formed by the Australian and State governments in 1964 to coordinate the development of forests in the interest of the community and to guide national programmes for the production, utilization and conservation of Australian forests. Its strategy document (AFC, 1986) sets out 'important basic principles and goals associated with the management of Australia's forests'. It consists largely of exhortatory statements and has no legal status.

Policy

Major issues have arisen, such as the destruction of indigenous forest, a topic which is argued most keenly where the income generated from logging is expected to be low. Otherwise issues which have become important matters within departments have been of an essentially tactical nature. The sensitivity of the actions of forest departments' to political pressures is remarkable and the attitude of those responsible for the direction of forestry in States is highly pragmatic. Such debates are, however, carried on within the existing mould with a public department managing lands an obvious target for local pressure. The massive changes of recent years in attitudes towards the ownership of forest resources seen in Britain, Sweden and New Zealand have not yet been fully recognized, let alone assimilated, in Australia.

In Western Australia, strategies for management of public lands have been developed following widespread public consultation (CALM, 1987). Management plans for regions translate these strategies for timber and for conservation and recreation into aims and, where judged practicable and appropriate, quantitative programmes. The dominant concern is with the use of the public forest. Private forestry is undertaken on a small scale compared with the activities of wood-using industry which owns substantial areas of plantations in Victoria, Queensland and New South Wales. Schemes for the encouragement of private planting are not established and maintained with the same continuity as is common in western Europe. A fundamental difference is that there is no direct control over private forest (Macdonald, 1989).

Implementation

A common technique for participation by private landowners in afforestation is the form of joint venture known as sharefarming (CALM, undated and 1992 describe schemes in Western Australia; Anon., 1988 sets out a scheme for Victoria). These schemes are similar to the reverse mortgages discussed in Britain in the early 1980s. They involve an agreement between the State and a landowner whereby the latter provides suitable land on which the forest department establishes and manages a plantation. In return the owner receives a predetermined annuity, indexed for inflation, plus a predetermined share of the revenue from final felling. The schemes are flexible, and the risk, though less than that with outright purchase of land, is largely borne by the State.

In Tasmania (Forestry Commission Tasmania, 1980) an 'annuity' scheme provided for a loan by the State to the owner who planted and received an annuity, repayment of the accumulated sum of these being made out of the value of the final crop. It is understood that this scheme has lapsed owing to lack of interest (Malajczuk, personal communication). A number of straightforward grants have also been offered in States. In Tasmania (Forestry Commission Tasmania, undated) these included provision of 40–50% of costs up to specified maxima for eucalypt planting and plant production, grants for amenity and erosion control and loans for pine planting.

A Special Case of Environmental Forestry: Water Use by Trees

One scheme is recorded because of the novelty of its purpose. Western Australia has a moderate rainfall in its south western corner and the area is noted for its highly successful sustained management of the native karri (*Eucalyptus diversicolor*) and jarrah (*E.marginata*). Tree clearance is prohibited in certain areas, and a scheme has been introduced (CALM, 1992) to afforest with *Eucalyptus globulus* on a pulpwood rotation of 20 years. This is a multipurpose scheme providing wood, shelter, erosion control and an increase in the loss of water by evapotranspiration. The reason for incorporating this last aim is salination. It is believed that rain and dry deposition introduce salt to the region's soils which has been transferred to the atmosphere from oceanic sprays. As Borg *et al.* (1988) explain, the upper 2–4 m of the soil are generally well leached, but substantial amounts of salt have accumulated at greater depths during the last few thousand years. With forest clearance in favour of grazing, groundwater levels rise, thus mobilizing some of the salt. This has then entered streams the salinity of which may rise if the increase in salt discharge is not balanced by an increase in runoff. The degree of salination may be high enough to be deleterious to human health, and it has been found necessary to use desalinated water in watering sensitive plants in the Botanic Gardens of Kings Park, Perth. Such treatment is expensive and the only alternative is mitigation of the problem at source. One method is to use *E. globulus* as a pump. The planting scheme requires a deed to be written setting out the contract under which the department provides the finance for crop establishment and management with an agreed figure for sharing of the crop at harvest. The farmer is encouraged to undertake the actual silvicultural work as a contractor, thus earning income at the time rather than only from a share of the final felling.

Observations on the Public Support of Private Forestry in Australia

The economics of wood supply have not reached the point in much of Australia where strong promotion of private participation in forestry is necessary. Government policy towards private forest management is accordingly mainly *laissez faire*, attention being confined to technical advice with a number of *ad hoc* schemes of assistance. The Western Australian eucalypt planting scheme is an exception, although it is conceivable that the treatment could be adopted in other States. There remains the perennial question of principle concerning the necessity for State involvement with forestry. Although there is no strong pressure from existing or intending timber growers to press for higher government encouragement, the interests of owners as well as conservation groups (Smith, 1989) are clearly best served by maintaining an accountable forest department. There remain marked disagreements, strongly voiced in the community, over objectives, ranging from preservation at one extreme to continued logging followed by intensive silviculture at the other. It appears likely that new initiatives relating to private participation in forest management will only come about when the aims of forestry as a whole have been clarified.

II. NEW ZEALAND

Background

Five facts are fundamental to an appreciation of the forestry position of New Zealand:

1. Forests cover 27% of the country, the bulk being the vestiges of the indigenous forest of which 6.2m ha remain: these have high conservation significance.
2. Plantations amount to 1.24m ha and supply all but 0.2m m^3 of the total cut in 1990 of 12.1m m^3.
3. The population is small at 3.4 million and, in common with all small economies, external dependence on the supply of goods is high; this in turn implies a high demand for foreign exchange and in this connection the export earnings of the forest products' industry, which account for 10% of the country's total, make an important contribution.
4. State forest holdings of plantations are minimal following the second set of sales of early 1992.

5. Financial rates of return on capital invested in plantations are as high, and investments as secure, as in forests anywhere in the world.

The fact that state forests can be sold provides no evidence of either the financial or social profitability of afforestation (cf. the case of Britain) since the market will decide the value of a going concern forest in the light of expected cash returns including any grants that may be receivable. Although the returns from 30 to 35-year rotations are still long delayed by comparison with other commercial and industrial investments, the prospects for the profitable management of a continually expanding resource are clearly acceptable to investors, a point already proved by the growth of New Zealand Forest Products Ltd a generation before. It is this feature which so sharply distinguishes New Zealand's plantations from those of much of western Europe.

Administration

The New Zealand Forest Service was dissolved in 1987, ownership and control of the state commercial plantations passing to New Zealand Forestry Corporation Ltd. Many forests had been sold by 1991 with a 30% slice remaining in the hands of the state-owned Timberlands New Zealand Ltd, earmarked for sale in 1992. Forest authority functions were taken over by the Ministry of Forestry. Its responsibilities include the provision of policy advice to government and of resources for research into forestry and forest products, regulatory activities and the administration of the residual Forestry Encouragement Grants and Loans.

The New Zealand Forest Owners' Association with a membership of 50 accounts for over 85% of the plantation area (Shirley, personal communication) and members include the big battalions of those older firms, or their successors following the reshaping in the aftermath of the sale of state forests, plus the new purchasers whose interests are also dominantly in wood-using or wood-exporting industries. It is estimated that some 150,000 ha are in the hands of 5000 small owners. The New Zealand Farm Forestry Association has 4500 members most of whose plantations are in small blocks worked in connection with their farms.

Policy

Ownership, protection and management of the indigenous forest have been effectively isolated from the operation of plantations, first, by transferring 4.9m ha from the management of the former Forest Service and Department of Lands and Survey to the newly formed Department of Conservation, and second, by establishing a policy for the protection and sustainable management of the remaining 1.3m ha held on government or private, half of it Maori, tenure. The continuing sensitivities which characterize discussion of the treatment of indigenous beech and podocarp forest particularly mean that there is little expectation of any change in the wood-producing capability of this resource, although Maori owners and others recognize that locally significant employment and income may be derived from its regeneration, enhancement and the ensuing increased production.

There is one intensely active front of forestry action in the country, namely that concerning the expansion, management and exploitation of plantations. The official view on the government's past and present role in plantation management is clearly expressed in Ministry of Forestry speech notes dealing with 'The forestry reform process' (1991a) from which the following extracts are taken verbatim:

1. The state did have a valid and valuable role in initiating large scale commercial plantations of a hitherto untried species;
2. Large scale plantation forests are a purely commercial investment and it is logical that government exit from the ownership role in favour of the private sector once early development risks had been overcome;
3. The multiple objective management role of the New Zealand Forest Service ultimately left it in a 'no win' situation and left everyone with a level of dissatisfaction;
4. New Zealand was able to successfully carry out the forestry reform process because it had sufficient mature plantation forests to service local industry and export markets;
5. Government retains a key role in administering and protecting conservation values;
6. Government retains a key role in the areas of forestry research, provision of protection and regulatory services, and in ensuring that the forestry sector is fairly treated in relationship to other industries and forms of land use.

Rarely can a change of policy have been so clearly, if laconically, described. The no-nonsense attitude is refreshing for its candour but many issues are raised by these remarks. Among them are the question of the attitude that would be taken by government today to a pump priming exercise comparable to that undertaken by its predecessors in planting *Pinus radiata*. The assertion on continued public support of research is also significant.

It is entirely consistent with the current philosophy of allowing market considerations free play that private owners are not obliged to restock after felling unless this is required by local authorities for soil conservation reasons. However, no doubt in recognition of the implications for maintenance of supply to mills, the second tranche of sales of former state forests, that of Timberlands New Zealand, required successful purchasers to replant unless they could prove a more productive land use.

A public discussion document (Ministry of Forestry, 1990a) followed the government decision to develop a forest policy. In support of this intention, it is of interest to note that the Ministry prayed in aid the contentions of Westoby (1989) on why a declared forest policy was desirable. These were, in brief, the dependence of an effective policy on a well-informed public, the importance of consensus in supporting continuity of policy in the long-term national interest and the value of a clear policy as a guide to a variety of managers, public and private sectors, in their day-to-day management. The wonder is that the new policy has not emerged earlier. Alternatively it may be argued whether there is any need for a policy in a country which has taken the radical privatization steps of New Zealand. However, two important concerns are the public reaction to plantations (reflected in the discussion paper) and fears about the treatment of the conservationists' dream, the broadly untouchable indigenous forest (Wilson, 1992, illustrates the continuing fascination with this resource).

The generally 'hands-off' attitude of government which is welcomed by commercial plantation owners, is in contrast to the concerns over the indigenous forest. Increasingly, in the absence of stronger national legislation, local authorities may decide to use their powers under the Resources Management Act to control or prohibit clearance of indigenous forest. Legislation to protect and manage such forest is still under discussion. Apart from the decisions on taxation and extension services referred to below, little by way of direct government intervention has altered to affect the non-industrial private forest owner. On the provision of unmarketed benefits, the 1990 discussion paper included a statement to the effect that forest owners may impose costs, possibly of an intangible nature (presumably social losses in such fields as landscape and wildlife conservation), and they may provide benefits such as recreation opportunities, but it was considered that 'It may not be productive to attempt to formalise such costs and benefits in the New Zealand setting'. This statement has to be set in the context of a country in which land issues, particularly with regard to Maori interests, remain sensitive.

Implementation

Against this background of a free market attitude towards plantation forestry it is not surprising to find that government offers minimal Exchequer support to plantation owners. Thus, in 1991 and after some vacillation, the government provided that the costs of plantation establishment and maintenance should be fully deductible against other income for tax purposes (Ministry of Forestry, 1991b). However, only 5% of the cost of land clearance is deductible, in contrast to the 100% which applies to clearance for grazing (MacBride, personal communication). The accrual system of offsetting expenses against income is as Saudrey and Reynolds (1990) point out, too complicated to operate satisfactorily in forestry. The tax rate on forestry net revenue is 35%. The Ministry no longer offers a consultancy service. Instead it offers a general information service on forestry and generally acts in a promotional role, for example by arranging investment seminars. The Ministry has announced its willingness to act as a facilitator for joint ventures between investors and landowners while remaining wholly aloof from any financial involvement.

It is noteworthy that continued public financing of research on forestry and forest products is to be maintained. A new development in this field was the establishment in 1990 of a Forest and Forest Products Research Organization (FAFPRO) with four component boards covering forest technology, harvesting, wood technology and pulp and paper, plus a separate Forestry and Wildlife Ecosystems Research Coordinating Board (Forest Research Institute, 1992). The aims of these organizations are to develop coordinated research programmes, to coordinate the securing of funds and the promotion of research.

Commentary on Public Policy on Private Forestry in New Zealand

Apart from soil conservation and some other residual obligations of forest owners, obligations that would be accepted as normal for responsible landowners in any country, the New Zealand position in relation to private forestry appears at first sight to come as close as possible to the concept of freedom exemplified by the sweet shop analogy referred to in Chapter 2.

Such a view would, however, be naive. Timber-marketing arrangements for landowners remain critical determinants of their continued economic health. The pressure on firms to be aggressive in finding and keeping new markets as wood supplies rise (they are forecast to increase some 6% per year over the decade 1993–2003) demands that purchasing be as keen as possible. The exercise of market power must constitute a concern where a single company dominates the roundwood market. Under these conditions, continued existence of log export markets is essential. Also, the owners' position is constrained by the diminished prospects for agriculture following the removal of agricultural subsidies in 1989. Thirdly, although there is nor the same level of interest in landscape, the kind of concern expressed so vehemently in Britain about the visual condition of the countryside if intensive agriculture ceased will no doubt be heard in New Zealand if plantations are allowed for any reason to decline.

It is clearly right that 'Forestry's importance in protecting and enhancing the environment will also require increasing attention from a forestry agency' is among the functions listed in Appendix 2 of Ministry of Forestry (1990b). The statement might be enlarged to note the need for the conservation of non-industrial private forest owners whose position is not unequivocally one that is more favourable because of the new disposition of industry forests and wood-using industry. A continuing programme of new planting of the order of 50,000 ha year^{-1} was expected to be undertaken after the end of the depression of 1990. Achievement of this target will be partly aided by the decline of livestock farming following the total removal of farming subsidies. A recent trend (MacBride, personal communication) has been for businessmen to form syndicates with which to buy plantable land or to form a joint venture partnership with the land owner.

III. EASTERN EUROPE AND RUSSIA

The massive political changes in the countries of eastern Europe and the former USSR imply a number of changes for forestry which are of wide interest for several reasons. Many of the new governments installed since 1989 strongly favour a market economy system. Several have stated a policy on the private ownership of forests and it is this aspect which is reported on here.

Land has proved to be a politically attractive asset to privatize and yet there is often little demonstrable by way of an economic rationale for such action. The reasons for the change of title are manifold, ranging from the ideal of restoring property expropriated at the time of the communist takeover to the families of the original owner, to the political desirability of developing a pluralistic society and organization of production. Unfortunately, it is not clear that in all cases the same urgency has applied to the consideration of management post-privatization. In some countries at least, the privatization of all or part of forest-based industry, which appears to have a much stronger economic rationale, lags behind that of the forest.

The redrawing of frontiers in eastern and central Europe, so frequent that a man born in Transylvania, for example, could well have found himself a national of five countries between 1914 and 1945, makes the use of statistics for different points in time hazardous. However, some consideration of changes is necessary in the absence of statements of the areas of private forest immediately before the communist take over of these countries in the middle to late 1940s. There were substantial changes in the boundaries of Poland and to a smaller extent Czechoslovakia from before the Second World War to 1945. There were only insignificant changes in the territories of eastern European states after 1945. Reference is therefore made back to the pre-war position since figures on the proportion of private forests are often available only for that time. Table 15.1 shows land and forest areas and the proportion of forest held privately in the inter-war period and in 1990. The coverage is of all eastern European countries. Major programmes of new planting, implying afforestation in Hungary and Bulgaria, and replacement of scrub or poor coppice incapable of timber

Table 15.1. Total and private forest areas in East European countries.

Country	Before 1939[a]	Total land	Total forest (m ha)	Private forest (m ha)	% of all forest	Total land in 1990[b] (m ha)	Closed forest[b] (m ha)
Albania	1934		0.99	0.063	6.4	2.78	1.05
Bulgaria	1933	10.9	2.97	0.542	18.2	11.0	3.39
Czechoslovakia	1930	15.6	4.60	2.96	64.3	12.5	4.49
Hungary	1934	9.1	1.17	0.86	72.9	9.1	1.68
pre-Poland	1936	39.1	8.35	5.34	63.9	30.4	8.67
Romania	1929	23.7	7.13	3.30	46.3	23.0	6.19
Yugoslavia[c]	*ca.* 1947	25.7	8.40	2.13	25.3	25.5	8.37

[a] Troup (1938) quoting figures published by the Economic Committee of the League of Nations in 1932 shows broadly similar figures to those indicated except that the proportions of private forest were somewhat higher for Poland and lower for Romania.
[b] ECE/FAO (1992). In addition to the closed forest area shown, the following areas of other wooded land occur: Albania 0.4m ha, Romania 0.08m ha, Yugoslavia 1.1m ha.
[c] FAO (1957).
Sources: before 1939 – International Institute of Agriculture (1936); 1990 – ECE/FAO (1992).

production in Bulgaria, increased the areas of forest in these two countries substantially. Of the remaining countries, Romania alone shows a marked loss of area.

Summary on Former Areas of Private Forest in Eastern European Countries

The proportion of total forest area owned privately in the early 1940s ranged between 17% in Bulgaria to more than 70% in Hungary, with over 60% in Czechoslovakia and Poland. In the former Yugoslavia, the percentages of private forests before 1939 were: Macedonia and Montenegro 0, Bosnia-Herczegovina 11, Croatia 24, Serbia 33, Slovenia 60. The period between communist domination of the east European countries and the overthrow of these regimes was 40–44 years. Many of the owners of estates taken over by the state were still alive at the time of the restoration of democratic government. Exceptionally, in Poland, where private farms accounted for 80% of the agricultural area, 16% of the woodland also remained in private hands. But elsewhere 'social' ownership was the rule.

The methods of privatization of forest land proposed in 1992 are very diverse, as shown for four countries below. The emphasis in the discussion is on transfers to private individuals; the position of forests formerly held by the church, educational establishments, etc. is not discussed. Details of the post-war changes under the incoming communist regimes are derived from FAO (1957).

POLAND

The result of the 1939–1945 war was that Poland lost a large area in the east to the USSR and gained a smaller area from Germany, the net loss of territory being more than 20%. However, the area of forest became slightly larger. Much woodland is associated with farms. The communist government expropriated 9m ha of arable land, consisting of 6.8m ha belonging to Germans and other minorities, and 2.2m ha of private estates. This was distributed thus: 5.8m ha to private farmers, 3.2m ha to collective farms. In 1953 cooperative farms accounted for only 7% of the farmed area.

In 1990, the area of private forests was 1,519,700 ha or 17.5% of the total forest area of 8,693,900 ha. There were about 1,600,000 individual owners. It is intended that certain of the forests nationalized in the 1950s, namely properties of not more than 25 ha, will be privatized. Larger areas which had been nationalized may be not be passed to former owners, but compensation made in the form of bonds.

Poland plans to afforest some 700,000 ha of the poorest agricultural land partly through private planting, partly by the state following purchase of farmland, so that the area of forest should increase to 30% of the land area, i.e. 9,380,000 ha. Of this, private forest would amount to about 20% or approximately 1,900,000 ha (Zajac, personal communication).

CZECHOSLOVAKIA

In 1945, forests belonging to Germans and 'collaborators' were expropriated. Next, all private forests larger than 100 ha were made state forests, and from 1947 to 1949 all forests over 50 ha, totalling 680,000 ha or some 16%, became state or communal forest. A further 30,000 ha were transferred to state forest ownership as a result of expropriation of farms over 50 ha. By 1955, 517,000 ha (12%) remained private, the great bulk, 486,000 ha, being properties of less than 10 ha. An interesting feature of the forest law of 1948 was that nurseries were required for communal forests of over 100 ha.

Czech Republic

Forest land amounted to 2,629,000 ha or 33.3% of land area in 1992. The ownership of forest in late 1991 was divided as follows:

- state and other public organizations 96%
- agricultural cooperatives 4%
- private 0.05%

The principle underlying the bulk of the intended privatization of state forests is that of restoration of property to the municipalities, private persons and the churches and religious communities (Rybniček, personal communication). The area concerned lies between 35 and 50% of the forest area, i.e. 800,000–1,000,000 ha. The area of state forest remaining, 50–65%, will be substantially higher than the 17% in the same area of the Czech Republic before 1939.

Slovak Republic[1]

In 1991, the forest area was 1,834,800 ha. A 1991 law enables the handing over of 765,300 ha, i.e. 41.7%, to the following categories of owner (Svitok, personal communication) (areas in ha):

• private	309,600
• company	416,900
• agricultural establishments	24,600
• other	14,100

It has been proposed that forests of local authorities (173,900 ha) and the church (55,500 ha), totalling a further 229,400 ha or 12.5%, should be privatized by means of a separate act. If implemented, the total number of owners will be approximately 300,000. It is intended that, because of the very small scale of many individual ownerships, voluntary cooperatives will be formed to manage these woodlots.

Hungary

In 1938, state forests accounted for only 5.2% of the total forest area, and 17.4% belonged to municipalities and communes, 14% to churches, 5.4% to foundations and corporations, 44% to individuals and 14% to trusts (fideicommis). Expropriation, starting with the large estates in 1945, led to the proportion of private forest falling to 8.3% by 1955.

Although by 1980, i.e. 9 years before the change of regime, Hungary had moved to a situation where market forces were allowed to guide prices to a substantial extent, changes in the ownership of assets have been small. It is intended to privatize two types of forest, first woods now forming part of agricultural cooperatives, and second, areas of state forest previously belonging to private owners. It is expected that as a result 40% of the total forest area will be privatized (Gerely, personal communication).

[1] Following the creation of Slovakia as a separate state, it is expected that changes in land ownership will be slowed.

Bulgaria

Before the Second World War, all forests, both state and communal, were overexploited; peasants often removed material from the nearby state forest, this being regarded as complementary to the farm. The area of formally private forest, 18%, was nationalized in 1947 and at the same time an immense programme of reafforestation and conversion of coppice was begun.

Following an interregnum, a new government of a liberal hue took office in December 1991 followed by a return to a more Marxist administration in 1992. There has been no strong tradition of private forest ownership in the country and no plans have been announced for privatization of state forest.

Russia

By far the greatest part of the former USSR's area of forest lies within the Russian Federation. Before the 1917 revolution, private forest accounted for 23.8% of the total area of European Russian forest, and peasant holdings to a further 2.8% (Zon and Sparhawk, 1923). The Siberian forests amounted to over 400m ha according to the same authority, two and a half times that of European Russia, but the area in private hands was only 4%.

A new Forest Code submitted to parliament refers to all forests as being federal property (Moiseev, personal communication). However, certain autonomous republics and regions favour municipalization of some state forests.

Commentary on the East European Scene

Whether forests are to be privatized or not, it is arguable that a most significant change, with potentially far greater economic repercussions, will result from private ownership of wood-using industry. In the past, state-owned industries were highly inefficient, heavily subsidized and paid very low prices for roundwood. This had a baleful impact on forest management, especially thinning. Privatization of this sector is likely to create strong pressure to increase the cut and manufacture of forest products for export. Roundwood prices on the domestic market are adjusting to world prices. In the process, a difficult choice has to be made on the maintenance of flows of roundwood and wood products to the domestic market while still encouraging exports. There are great incentives and scope for large increases in roundwood production. Forest authorities in these countries carry the heavy responsibility of appreciating and reacting appropriately to the new situation created by market realities which will increasingly drive the actions of the new forest (and mill) owners.

REFERENCES

Anon. (1988) Sharefarming: new opportunities for land owners. *Australian Forest Grower*, Winter, 26.

Australian Forestry Council (AFC) (1986) *National Forest Strategy for Australia*. Canberra.

Borg, H., Stoneman, G.L. and Ward, C.G. (1988) The effect of logging on groundwater, streamflow and stream salinity in the southern forest of Western Australia. *Journal of Hydrology* 99(3/4), 253–270.

CALM (Department of Conservation and Land Management, Western Australia) (undated) *Softwood sharefarming*. Como, WA.

CALM (1987) *Submissions on Management Plans for North, Central and Southern Forest Regions*. Como, WA.

CALM (1992) *Grow Trees for Profit: Eucalypt Timberbelt Sharefarming 1992*. Como, WA.

ECE/FAO (1992) *The Forest Resources of Temperate Zones: Results of the UN-ECE/FAO 1990 Forest Resource Assessment*. Geneva.

FAO (1957) [*The Forestry World, Vol. 1, Europe and the USSR*.] FAO, Rome.

Forestry Commission Tasmania (1980) *Pine Plantation Annuity Scheme*. Hobart.

Forestry Commission Tasmania (undated) *The Private Forestry Division*. Hobart.

Forest Research Institute (1992) *Report July 1990–June 1991*. Ministry of Forestry, Wellington.

International Institute of Agriculture (1936) *International Yearbook of Forestry Statistics*, Vol. 1, *Europe and USSR*. Rome.

Macdonald, G.T. (1989) Rural land use planning in Australia. In: Cloke, P.J. (ed.), *Rural Land Use Planning in Developed Nations*. Unwin Hyman, London.

Ministry of Forestry (1990a) *A Forest Policy for New Zealand: a Public Discussion Paper*. Wellington.

Ministry of Forestry (1990b) *Post Election Briefing*. Wellington.

Ministry of Forestry (1991a) *An Outline of State Encouragement and Ownership of Commercial Afforestation in N.Z.* Wellington.

Ministry of Forestry (1991b) *Trees and Timber – Forestry Taxation*. Wellington.

Saudrey, R. and Reynolds, R. (1990) *Farming Without Subsidies: New Zealand's Recent Experience. A MAF Policy Services Report*. GP Books.

Smith, R.P. (1989) *Timber Production: the Grower's Viewpoint*. Proceedings National Agricultural Outlook Conference, Session 4.

Troup, R.S. (1938) *Forestry and State Control*. Clarendon Press, Oxford.

Westoby, J. (1989) *Introduction to World Forestry*. Basil Blackwell, Oxford.

Wilson, G.A. (1992) A survey of attitudes of landholders to native forest on farmland. *Journal of Environmental Management* 34, 117–136.

Zon, R. and Sparhawk, W.N. (1923) *Forest Resources of the World*, Vol. 1. McGraw-Hill, New York.

16

CURRENT POLICY ISSUES IN WESTERN EUROPE

RECAPITULATION OF PURPOSE

The first aim of this study is to identify policy issues which are either current or emerging in the countries reviewed, particularly those of western Europe, and to gauge their relevance to Britain. The countries considered in Chapters 6–14 were chosen on the basis that their social, cultural and economic characteristics were broadly similar to those of Britain. It is hardly surprising to find that no radically different issues from those current in Britain are discussed. The similarity of the spectrum of themes that concern people and politicians in west European countries is marked, regardless of whether they are wood rich, have a particularly strong forestry tradition, are keen on hunting, or none of these, but the emphasis given to each issue naturally differs between countries as does the extent to which policy implementation has developed in relation to particular issues.

THE ASSESSMENT OF ISSUES

Before it is possible to formulate a policy that has any pretence to rationality, it is essential to identify the issues that concern people, to understand how and, as far as possible, why they arise, and to understand how they interact. The following discussion of issues concentrates on the basis for the claims made by different groups and the methods to be adopted to weigh these in the balance.

Old pragmatic ways of resolving conflicts are increasingly questioned. One pervasive issue is man's role in influencing the environment. This is regarded as being sufficiently important to justify separate consideration. A second general consideration is the extent to which public reference and participation is made a part of the discussion of policy issues.

Environment, ethics and economics

Important questions of environmental ethics lie behind many of today's forestry issues. The extent to which environmental economics should determine policy, as argued here, is one of them. Useful reviews indicating different perspectives in these areas are provided by Alston (1990), Anderson and Leal (1990), Grove-White (1991), Norton (1991) and Pearce *et al.* (1989). A fundamental concern is with the extent to which weight should be given to such a goal as the protection of nature for its own sake, that is outside any utilitarian philosophy or even one which recognizes the essentially anthropocentric idea of 'existence value' (that is the value a person associates with knowing that an asset exists even though he will never be able directly to experience its enjoyment). Nature conservationists come closest to acting as if in pursuit of some such ethic and it is troublesome to record that these large issues are not more widely discussed in forestry circles in Britain.

It is not particularly profitable to speculate on the consequences for forestry policy in Britain, or Europe for that matter, of a radically different approach from that implied by western society's value system. For example, a society which embraced a totally different view from the Judaeo-Christian tradition of adapting nature to man's purpose would no doubt have very different attitudes to native forest clearance, the use of exotics, genetic manipulation and interference with water chemistry from those commonly expounded and demonstrated in Britain today. Beatley (1991) supplies a good example of the absolutist position founded on ethical principles; unfortunately the basis for many of his assertions is quite unclear. Thus, the view that 'ethical land use policy sustains and protects natural ecosystems; ethics requires a small human "footprint"' leaves little room for man, and none for economic growth. Although he recognizes that there may be conflicts between individual principles and that a code of conduct is required to make the scheme operational, Beatley does not show how such a code can be derived. Nevertheless the example of this approach is useful to remind policy makers that value systems do exist which are different from those to which a majority in this country, perhaps too unthinkingly, subscribe. Such ethical concepts may well explain the attitudes of those who express concern about the use of forests in continents that have been less dominated by man's action than Europe but which are now increasingly subject to those pressures. Recognition of these attitudes must help prepare the policy debate at home for the possible emergence of new aims for forestry.

There are in practice many issues to resolve within the traditional European approach to the use of trees for man's ends. Two illustrations of the point may be made. The first stems from the absolutist position assumed by stating man's right (*sic*) to clean air and clean water. Cleanliness is a rather well established convention in relation to drinking water and recent EC directives require ever higher standards to be met. But the principle may be extended to the quality of water entering watercourses and may come to be applied, for example, to strict control of road building in forests where soil disturbance may cause high turbidity even though the cost of remedial action by the water supply authority would be lower than that of avoiding the impact on the stream or river. Or again, liming of water source areas to offset the effects of acidic inputs captured by the tree canopy (Forestry Commission, 1991) may be questioned on the 'precautionary' principle that the impacts of such liming cannot be, or have not been, predicted. A second example centres on the application of economics in circumstances such as those noted above concerning turbidity. Legislation on the environment which establishes standards may well go beyond the economically efficient point at which marginal social cost of, for example, pollution equals marginal benefit. It is possible that this behaviour simply means that the economic 'optimum' has been derived in ignorance of the full facts. Only recently have economists defined the concepts of option value and existence value. It is just possible that these considerations underlie legislators' decisions and that incorporation of these values into the calculation would justify the standards set. In a field such as human rights, society establishes what constitutes acceptable behaviour and legislates accordingly. But it would be wrong to conclude that standards laid down by law are in general the appropriate way to decide policy in all fields and there appears to be no reason for distinguishing environmental matters as ones where certain absolutes apply.

It is argued here that the rational approach to policy is, as far as possible, to ascribe a money value to the output in each attribute, and to debate the implications of acceptance of the mix of outputs which arises from the calculus using such valuations[1]. If the result is acceptable, policy may then be developed as necessary in the sense of being refined for particular situations (for example of region or ownership). Secondly the values derived become essential management tools replacing arbitrary guidelines and allowing the calculation and trade-off of gains in one output against losses in another.

As a general rule the conclusion of this review on the means used to arrive at new policy stances is that the main technique used in Europe is open, sometimes very lengthy, discussion. Readily quantifiable elements such as national income generated by forestry, regional labour income and employment, and export earnings are selectively used in support of arguments. Most non-market outputs, which are by no means always unquantifiable and

---

[1] Procedures such as goal programming cannot be regarded as alternatives, let alone panaceas although they aid clear thinking where outputs have not been evaluated and are therefore incommensurable.

incapable of being evaluated, are still handled by reference to the claims of pressure groups.

There are marked differences among countries in the emphases given to attempts to find a more critical approach. Thus Germany, the country in which objections to the form and intensity with which traditional silviculture has been practised are most strongly voiced, conducts next to no research on the valuation of non-market benefits in forestry, thus underlining the slight attention paid to economics in forest management in that country. Nordic countries have active programmes. There are programmes in various land-based fields in the United States concerning evaluation and wide use is made of planning tools to combine incommensurables. Britain has carried out and continues to undertake a considerable amount ofresearch into the valuation aspect. Major gaps however remain; for example, little work is conducted on the values to be ascribed to gains in employment and population in disadvantaged areas. Exchanges take place through bodies such as the International Union of Forestry Research Organizations (IUFRO) but a more coordinated approach here and in other fora (Opheim, 1991) could ensure speedier testing and application of ideas in a field of universal concern.

Most of the policy statements referred to in country chapters are not specific about the emphasis across owners. There is a risk of confusion about purpose where the agents of policy are not defined but in practice no example, with the exceptions of afforestation of farmland where the aim of policy is in any case clearly oriented towards a private farming industry, and provision of general public access where this is not an existing right, has been found where a national policy aim is not regarded as also applying to private forest owners. As noted in Chapter 2, a further dimension is introduced when policy towards private owners is considered. Most owners would prefer to rely on revenues derived from the market, and preferably several markets, than on those from government in the form of grants, largely on grounds of greater reliability of the market than the exchequer. This may mean that government has to accept that costly incentives are necessary if private growers are to be tempted to help with the achievement of certain social aims.

Public consultation

It is implicit in the pluralistic societies which characterize western Europe that decisions made by government and its officials have to be acceptable to the public. While the uncertainties surrounding the Maastricht Treaty reflect on governments' and parliaments' ability to reflect popular feeling, these institutions must remain at the heart of decision making. It is not difficult to see how public participation in the local implementation, as opposed to formulation, of policy may be arranged. The more difficult area concerns the way in which the broad lines of policy are delineated. Hummel *et al.* (1984) and Westoby (1989) emphasize the importance of ensuring that peoples' needs and wishes are respected in policy making. There is less discussion of the ways by which this may be achieved. Ellefson (1992) points out that direct involvement of the general public in the establishment of policies is a rarity, but the public's passive involvement can at times be awakened by deep-seated concern over certain issues. As has been pointed out elsewhere in this study, those setting the agenda for policy formulation remain unskilled in identifying which the 'certain issues' are.

Classification of Issues

The following sections discuss the various issues identified as important in one or more of the countries visited. Emphasis is placed on attempts to define and measure the target output, without which management is rudderless and prospects of valuing the particular output are dim. The classification adopted here considers issues on the output side separately from matters concerned mainly with inputs. Principal non-wood uses are referred to first.

Outputs

Multipurpose forests and the demand for new forms of silviculture

By themselves terms like 'multipurpose', 'multiple use' and 'environmentally friendly' have no operational significance. Where such words are bandied about it is hardly surprising that a great deal of effort is devoted to propaganda by both sides of the inevitable divide between those whose income is affected by these claims and those whose interest is broadly in the achievement of some aesthetic, ethical or other non-monetary objective. It is difficult to avoid mention of the concepts however vague their meaning: the growing point of forestry debate continues to be the scale and kind of adjustment desired to meet the demands for more beautiful, more varied and more wildlife-rich forests. Thus, a more friendly forest is the basis of the moves among policy makers in Norway and Sweden to ensure that private owners provide a 'richer forest'. The concept is already a firm part of much private forestry practice: indeed it would be odd were it not so since so many owners are individuals or members of family trusts living close to their woods.

Many of the demands of those wishing to see more friendly operations practised are founded in ignorance, such as concerns over effects of localized application of herbicides two seasons per rotation on the assumption, presumably, that their use is likely to eradicate all wild vegetation in the locality. Martres (1991) provides a catalogue of the types of claim made by those whom he charitably suggests may simply be ignorant, rather than by those pursuing some quite different objective under the guise of a concern with nature conservation. In some countries, such as Germany, the pressure is very much a single-issue affair, with environmentally inclined groups minimizing the losses of income to owners and labour in both forestry and the wood-using industry, as well as the impact on the maintenance of employment in remote areas. In other countries, the criticism is more measured and may relate to the unacceptable rate of change in forest practice, especially harvesting, as in Norway, and the desire to return to more familiar patterns and processes in silviculture.

In Germany pressures to adopt 'near-to-nature' forestry continue to grow. There must, however, be major reservations about the validity of leaving all to nature in montane areas. In the United States, strong claims have been made for the merits of the 'New Forestry', a form of silviculture for use in regenerating and maintaining the old-growth forests of the north-west Pacific coast. The name has been given to a less intensive exploitation than that associated with large-scale clear cutting followed by replanting. The system leaves some trees growing, retains dead trees and ensures that material is left on the forest floor. The claims made by one of its advocates, Franklin (1989), have been aired widely (Dayton, 1990; Gregg, 1991) but as with most discussions on silvicultural systems the inability to offer experimental evidence means that protagonists depend mainly on assertion. It seems unlikely that any continental European forester working in the mountains would accept Franklin's claim that leaving dead trees and unharvested logs provides good protection of hillsides from water erosion.

But there is one aspect of the argument which may well have force. This is that the reaction against large-scale clear felling may find sympathetic ears in Britain also (Britton, personal communication). Continuous cover forestry along with other systems and operational techniques that fit ideals of working which are claimed to be 'close to nature' form a potent brew which foresters do well to heed. The fact that seeding of many species in Britain is not so regular or heavy as with the predominantly native trees of America and continental Europe is not a sufficient excuse for ignoring the virtues of low-intensity silviculture exploiting natural regeneration and working in much smaller coupes. Again, the fact that systems work better with shade-bearers than light-demanders and that the central European forester has a wide range of such species at his disposal is not sufficient reason to consider that such systems cannot be operated in Britain. The subject of the choice of silvicultural system has simply not been considered carefully enough in the circumstances of private estates.

One further dimension of the debate over forms of silviculture may be identified under the heading of sustainability (see Pearce *et al.*, 1989 for a discussion of the theme). Assuming that soil erosion is not at issue, the main concerns can only be continued health of the forest and absence of nutrient depletion. Healthy condition is one of the targets of those arguing for a return to more natural forestry in central Europe who find their greatest support in Germany. In view of the 1980s' hysteria over prospective decline of forests generally in that country it is not surprising to find the link between forest damage and the move to 'near to nature' forestry most strongly claimed there. Substantial public provisions for forest 'protection' have been made by aiding programmes of soil amelioration in Germany. In other countries, work on the ground has been limited to research and demonstration projects, although soil acidification is considered a potential threat to growth in southern Sweden. Nutrient depletion through intensive cultivation is another possibility. It seems likely to be a limitation on very few soils. If the effect is detected, pursuit of sustainability in its purest form would imply reversion to a less-demanding form of cropping or variation of the form of harvesting, for example retention of bark on site, or total cessation of harvesting. Intensive work such as that by the Forestry Commission and the Institute of Terrestrial Ecology at Beddgelert, North Wales, is necessary to detect any such effects.

It is not feasible to construct a catch-all mensuration for environmental friendliness. Instead individual components, some of which are noted below, must be specified, measured and evaluated and synergies and conflicts among the outputs concerned must be identified. A marked gap is the lack of good evidence on the forest management costs and wood yields of different systems: this is a technical problem calling for a combination of silviculture with timber mensuration which remains unresolved but should be soluble.

Wildlife conservation

Some of the goals that become labelled 'nature conservation' are so largely concerned with the visitor's enjoyment or potential enjoyment that it is appropriate to recognize them under recreation. This appears to be the main concern of the forest policy favoured in Flanders, for instance, and is clearly the reason in Britain for nature conservationists' protection of habitats famed for their butterflies. Indeed it is not unreasonable to suppose, as Aronson (1992) claims, that the motivation for scientists to preserve species and habitats is partly aesthetic.

Most European countries which have an active, usually governmental, nature conservation organization have adopted the kind of site description and designation adopted by the former Nature Conservancy Council (1989) in Britain. Many also have a hierarchy of degrees of protection of sites from national parks downward. A particular feature in countries which are forest-rich or have a high proportion of long-established woodland is that the forest is of high biodiversity interest and large areas have long been designated for special non-timber or low intensity timber production. Pressure is continuous for further designations in, for example, Germany and Sweden. 'Ancient woodland' is well established in Britain as an operational concept with clear management implications (Forestry Commission, 1984).

A crucial feature of these designations and the management constraints they impose is their elitist nature. In other fields, the appeal of the public forest manager is to the public and the preferences they show: it is their response to silvicultural practices, access to forests, recreational facilities, the distant view, and even to jobs provided which is heard and noted. The expert in each of these fields may be on tap but he is not normally, as is the case so often with nature conservation, on top. This point of difference is a fundamental one. An example of potentially great significance to foresters illustrates the argument; this is the attention which may come to be given to a particular woodland species. In Washington and Oregon the arguments on banning the logging of old-growth timber on Forest Service lands are focused on the fate of the northern spotted owl. In both Norway and Sweden the comparable species is the white-backed woodpecker on which economic work using such techniques as contingent valuation is beginning[1]. It is interesting to speculate when and on what woodland species attention may first be concentrated in Great Britain and for what reason.

Two stages are involved in attempting a disciplined and more objective approach to weighing conservation aims in the balance. On quantification, the state of the art of measuring conservation output is primitive. A system is however evolving. Usher (1986) provides a useful assessment of a number of multidimensional quantitative methods and Usher *et al.* (1992) illustrate applications to the assessment of plant richness in farm woodland. The second step, evaluation, is even more critical. Even if it were possible to adopt a 'democratic' approach, it has to be remembered that consumer sovereignty is by no means accepted as the sole guide in all walks of life. It is arguably not reasonable to follow without question the guidance of monetary measures, whether derived from the market, assessed indirectly via behaviour or by inventing a contingent market; but the lack of any economic assessment based on people's preferences is grotesque and indefensible.

Recreational opportunity

Recreational use of forests is probably the biggest single element in the thinking of those who strive for a more 'friendly' forestry. Actual visits to forests

---

[1] Contingent valuation uses the device of a hypothetical market to deduce preference. For commentaries on applications of CV procedures, see Benson and Willis (1992), Navrud (1989), Pearce *et al.* (1989) and Winpenny (1991).

are, after all, the way in which people come to see what is going on in forests. In Norway, Sweden and Germany, rights of access are so free, and in other countries suchas Denmark so relatively free, that public access *qua* access in these countries is not an issue. Policy to satisfy recreation demands instead centres on the provision of particular (unpriced) attractions. But wherever the emphasis lies, most countries pursue policies of engaging private forest owners in the process of increasing opportunities for informal recreation. France is the main exception.

Many studies have been made which quantify the recreational use of forests. A notable example is Denmark's 'People and forest' programme, a line of study initiated by the work of Elers Koch (1978, 1980). A further advance is to derive ordinal measures of preference. Lee (1993) illustrates the use of carefully administered questions to elicit preferences from visitors in Britain. Useful planning guidance is given by census-type enquiries, but until monetary values are available discussion cannot advance far. Monetary evaluations have been carried out in Norway, Denmark, Britain, the United States and to a lesser extent The Netherlands. Applications of methodology continue to grow: a British example is provided by the work of Willis and Garrod (1991) who found, using the travel cost approach applied to an individual visiting the site in question, values remarkably lower than those derived by the alternative method which considers the travel cost for the population from each zone (Willis and Benson, 1989). Interest has been extended in the 1980s to the more challenging evaluation tasks associated with the concepts of option, existence and bequest value. While the evaluation of benefits progresses, the work is based on parameters derived empirically from visit numbers, data on origins, etc. Understanding of variation in these parameters is lacking.

Landscape

Policy objectives often implicitly and vaguely refer to landscape. Surprisingly little attention has been paid in recent years to the evaluation of the contribution of different forms and extents of forest to landscape enhancement. After a flowering of studies advancing methods of ranking landscape attractiveness in the 1960s and 1970s, little further work has been done, apart from that by Price (1978), either in Britain or continental Europe. The relevance of the subject is greater outside countrysides dominated by forest massifs such as those of central Europe and the Nordic countries where landscape values will usually be subsumed by the benefits enjoyed by recreation visitors. Elaboration and reassessment is needed in order to confirm the conceptions in use by landscape architects and to develop practical guidelines comparable to those for other non-market benefits.

Job provision and community integrity

Policies for the generation of employment as a basis for maintaining community integrity apply in Norrland in Sweden and in remoter parts of Norway. These have some force for private forest owners in Norway. In north Sweden the proportion of private forest is quite low but private owners are expected to make their contribution and special grants have been available, but are now declining, to encourage job creation. Special aid is given in Swiss cantons with remote communities. In Ireland, job provision in areas where farming is particularly disadvantaged is a fundamental part of policy and of great relevance to farmers and their families. In no other country among those surveyed is job provision ranked as important. The support for this function is correlated with the political complexion of government, but the general shift in political philosophy of the past 15 years has had a dramatic effect on the pursuit of this objective. Studies of the effectiveness of special forestry aid are few but it must be recognized that assessments of the value of assistance to local income and employment are best conducted across the whole economy of the area in question.

Although the subject is of importance only in limited areas of Britain, evaluation of the maintenance of community integrity deserves attention. As noted in Chapter 5, the distinct matter of the social cost of labour in areas with low employment opportunities is problematic. Because its relevance to investment appraisal is of such high potential importance, further work appears desirable.

Forest influences

This term is used to embrace the physical effects of forest cover on the surrounding countryside through effects on the hydrological cycle, including heat exchange, flooding, etc., soil retention, shelter, and similar physical effects. It is strongly held in Switzerland that the physical parameters in play and the impacts of particular silvicultural measures have been well known for many years. Not all countries are in this happy position and indeed experience concerning forests and water in Britain suggests that much more remains to be learned before foresters can be confident about the effects of their management on water yield and flood control in one country at least. There is also a tendency in this field to absolutist thinking, and to hold that the same goals should apply to private and public forests equally.

There are no difficulties in the mensuration of effects on water (for instance, volume yield in months below storage capacity and flood peak attributes) nor in evaluation.

The new type of influence arising from the capture by tree canopies of acidic atmospheric pollutants and their transmission to the soil and hence water courses has led to the adoption of guidelines (Forestry Commission, 1991) in Britain which may have serious impacts on the scope for extending forests. It is curious to find no worry expressed in continental Europe, as opposed to Scandinavia, over the downstream effects. Monitoring of forest stream quality, presumably as a precautionary measure, is now being undertaken in Ireland (Mulloy, personal communication). Although soil acidification has been noted in Scandinavia where trial programmes of liming have been started, it has become an important issue only in Germany where policy has attended to the supposed damage to soil and hence trees. Despite major efforts, there remains a clear need for objective quantification of chemical and biotic effects in this field. Little work has been carried out either on willingness of anglers to pay and the manner in which this changes with water acidification, or of the cost of remedial action in treatment plants. Quantification of any soil acidification effects arising on plants including trees is more difficult, partly owing to genetic variability of stands.

Carbon storage

There are many excellent publications available covering the field in general, notable among which are the reports of the Intergovernmental Panel on Climate Change (or IPCC, for a review of the 1990 cycle of which see Grayson, 1991), and, in individual countries, those from France (Anon., 1990a) and Norway (Interdepartementale Klimagruppen, 1991) are informative. Swiss (Anon., 1990b) and CEC (1991) reports set out the arguments on installing a tax on CO_2 emission clearly. Assessments of climate change effects, and hence the benefit of limiting additions of greenhouse gases, are now high on the research agenda. There is as yet no carbon storage policy in place in any of the countries surveyed, either in forestry or other sectors. The subject is at the stage of measurement and description and although studies abound, their normative significance is small. While the claimed virtues of forestry, afforestation and wood use are widely broadcast, little analytical work on the value of forestry activities has been carried out. Instead attention has focused on the effects of increased concentrations of CO_2 on tree growth and outside Europe on the impact of climate change itself on forest health.

It is important to grasp that there is little gain to the world from forestry activities unless either or both of two things happens; one is that oxidation to CO_2 is slowed, for example by reducing the wasteful burning of tropical forest and by favouring log conversion into longer-life products, and second that there is substitution of wood for materials which occasion a bigger release of CO_2 in their manufacture. There is, it is true, a slight bulge in the trend of growing stock, and hence carbon, over time occasioned by additional tree planting (afforestation, that is planting of bare ground) but this is of limited duration and value before price effects work to diminish the incentive to carry out and maintain such investment. None of these steps can be considered regardless of cost: there has to be a trade-off between carbon emission and the cost of achieving reductions. This leads naturally to the question of the possibility of bringing carbon emission/storage choices and effects into the market system by taxation and subsidization.

In Switzerland the carbon tax under discussion in 1991 was to be based on fossil fuel use. The same base has been adopted for surcharges in Norway and The Netherlands. The debate in the European Commission (CEC, 1991) on the form of a carbon tax has centred on two main issues; one is concern about the impact of any such tax on particularly fossil-fuel intensive industries, the other is the desirable balance between the proposed basis of charge between fuel use on the one hand and carbon dioxide emission on the other (the first leg being to encourage fuel efficiency directly). Given a carbon tax, it becomes appropriate to establish a carbon storage credit or subsidy.

Solberg *et al.* (1991) have calculated the increasing marginal cost of various silvicultural operations which influence the level of the growing stock. The impact on the world carbon flows is, however, not made clear, though this is a vital step before it is practicable to formulate strategy, since a reduction of cut in one period will lead Norwegian consumers to look elsewhere for the local wood foregone. Nevertheless, the study is indicative of the way in which a carbon subsidy would influence management. A study for the Swedish Ministry of Finance on substitution of nuclear power by wood energy investigates the merits of an increase in wood production and hence growing stock (Hultkrantz, personal communication). This implies a real gain in world carbon storage to the extent that fossil fuel demand would otherwise increase.

Fuller and more open discussion of the whole subject of carbon is taking place in other countries (see, for example, Palmer and Krupnick, 1991 on a study of the effect of changed prices on electricity generation in the US) than in Britain where Anderson (1991) and Pearce (1991) have laid useful groundwork. The mensuration is easy, but major efforts are required to compute changes in equilibrium prices, including forest products, throughout the economy as a result of the application of carbon taxes and subsidies across industrial countries and the subsequent impacts of changed log prices on desirable species choice and management regimes.

Hunting and sporting

Hunting raises peculiar difficulties, since although market measures of value to the landowner are readily available, attitudes towards the sport over and above money values appear to dominate actions in the forest. Forestry policy is generally silent on the matter of provision for sporting, and particularly absent by reference to private woods, although there are often highly specific laws and customs governing the activity and in some countries the forestry administration is responsible for hunting. In relation to wildlife protection, there are no marked conflicts in western European countries. The ban on hunting in the Canton of Geneva is, however, a significant development. There are aspects of research overseas on both the biological and management side that are noteworthy in the case of deer especially, species that are of growing importance in Britain because of their currently rapidly changing distributions and densities. It is commonly said in Europe that both sportsmen and foresters are concerned over the level of damage due to game; in the event little appears to be done to restrict populations. The lesson from the countries surveyed appears to be that the impact of sporting demand on other forest values provides an awful warning.

Studies of preferences, including use of contingent valuation methods, can assist in achieving better management of game. Mattson (1989) has shown how a better allocation of elk hunting among hunters would not only result in a higher value of elk hunting itself, but would achieve this, as well as a higher value of hunting of other game species, without increasing game populations. The data for quantification and evaluation of the effects of alternative strategies are increasingly available in Britain for species of major concern such as red and roe deer but have not been exploited for benefit assessment and policy guidance.

Wood

It is clear that a policy on wood production exists and that this extends to the private sector in all countries surveyed, but remarkably different cultural attitudes towards wood production apply across countries. France and Germany (former FRG) have extensive forests with annual cuts of 45–50m and 30m m^3 respectively, totalling in excess of the combined cut of Norway with 10m m^3 and Sweden with 60m m^3. The contrast between the two groups lies not so much in the attention given to wood supply as to the production economics of the activity. In the Nordic countries the maintenance of a competitive wood-using industry is an important aim. Wood production is also regarded as important in France where the anxiety to diminish imports is a noteworthy feature and much attention is paid in debate to the need to increase the competitiveness of wood supply to mills and to achieve more economic conversion, but performance in this field falls far short of that attained in the Nordic countries. The overt emphasis on wood production in Germany is not so marked; an indication is given by the fact that the 1975 Forest Act is remarkably shy about the use of the words 'wood' or 'timber'. The position on the supply price of roundwood in Switzerland borders on the disastrous and suggests that a searching look at the possibility of continuing containment of logging costs would be valuable.

At the same time, and remembering that private owners have a special interest in the one main revenue producer, timber, the surplus capacity in many countries' forest estates is of concern. The weakening of standing tree values which is world-wide and has been noted in evidence from Norway, Denmark and, most dramatically, Switzerland, is accelerating the promotion of a shift in policy from wood production to the provision of more non-market benefits. In Sweden, Hultkrantz and Wibe (1991) have proposed that, apart from ensuring continued forest cover through regulation of restocking and control of insect pests, the state should relinquish its control over private owners' silvicultural practices, leave these to market forces and instead lend support solely to the environmental services of the forest. One cannot but wonder how possible it would have been to argue thus in the mid-1960s, a period when prices were relatively high and from which many of the investments now producing so much wood date. It seems quite likely that the cobweb theorem (high prices in one period lead to increased supply in a later one and so to a price fall which in turn reduces investment, so limiting supply in the following period which causes prices to rise once more) may indeed apply in forestry on at least a regional scale. Some bodies, notably the pulp and paper industry (Asp, 1990; CEPI, 1992) have issued warnings on a possible 'deficit' in wood in Europe. Asp's paper is largely based on the extreme and unrealistic views of the International Institute for Applied Systems Analysis (IIASA) on the supposed impacts of air pollution on tree health and decline. CEPI (1992) forcefully deploys familiar arguments about uncertainties surrounding future supplies. In relation to wood as an energy source, Hultkrantz (1989) provides a useful generalized supply curve for bioenergy in a particular region of Sweden based on work by Häckner (1988), which shows that bark, peat and sawdust have the lowest unit costs of energy. Pulpwood at then current Swedish prices was 60% more costly as an energy source than peat from newly opened bogs, the next cheapest contender as a biofuel. This result confirms the continuing impression that major changes in energy prices are required before wood becomes a significant biofuel for use in industry.

Arnold (1991) has provided a comprehensive outlook statement for wood. The general conclusion is that unchanged real wood prices are likely for several decades. While some appreciation remains to be made about the long-term prospects for exports from the forests of the former USSR, now almost entirely within the Russian Federation, this survey and those of ECE/FAO in timber trends studies for Europe have salutary messages for those considering policies towards any increase in the scale of afforestation, including that of farmland.

Several countries have undertaken a pulse of new planting so adding to a region's wood production potential. A technical problem which promises to become increasingly important concerns the calculation of the most economic method of avoiding a hump in wood yield followed by a trough, taking into account trade and industry investment aspects.

The enthusiasm for the intensive production of wood biomass for fuel tends to display a moral fervour. The economic potential of such production may be greater for other uses of fibre; only Sweden appears willing to support the concept beyond the research stage.

INPUTS

Several inputs are elements of forest policy; some are of lasting concern such as that on labour's opportunity cost noted above, others are more transient.

Ownership structure

Privatization is moving rapidly in some former communist countries but more cautiously in others. Any lessons to be learned in this process are for the countries concerned rather than Britain. A feature of privatization in some eastern European countries is the concern with restitution, a process demonstrating a touching faith in the merits of private ownership even though the current beneficiary lacks the skills and often the objectives of his forebears. Privatization of state forests has moved apace in New Zealand and is likely to take place in Sweden. The earlier Swedish example of the exertion of market power by the large forest-owning industrial firms will be in mind in any change in ownership in that country. The method of disposal adopted there will be of the greatest interest to governments and private growers alike in other parts of Europe including Britain.

Size-class structure

Many countries sustain policies in forestry which discourage fragmentation. The sociopolitical reasons for the widespread dependence on inheritance taxation, that is taxation of beneficiaries rather than the estate passing, are clear, but the resulting incentive to disperse wealth is clearly unhelpful in the case of forests. Despite such pressure the small and generally declining proportion that forest value bears to a person's total assets usually allows sensible transfers to take place without occasioning damaging fragmentation from this cause. This appears to be true even in France where inheritance rules favour the splitting of property. With the emergence of contractors in most countries, less emphasis is placed on the goal of establishing larger ownerships. This trend probably reflects the current liberal philosophy which includes the goal of wider rather than more concentrated private ownership.

Management structure

All countries recognize that private forestry suffers from scale diseconomies and many legislate to capture the potentially large efficiency gains to be made through economizing in the use of labour, machines, supervision and planning through cooperation. The variety of forms of such cooperation and the emphases across sizes of estate is very great. The relevance of the institutions that occur must be viewed in the context of the demands of owners and the prospects for succcessful collaboration in Britain. Increasingly through the post-war years, the cooperative function has been overtaken by private foresters' dependence on contracting firms. There is little point in developing policy in this field if there are no good grounds for believing that one or other form of cooperative is likely to be supported. The guidance from agriculture would not suggest that British landowners, whether 'traditional', new absentee owners, or farmers carrying out small-scale planting, are likely to use the cooperative route.

Technology of field operations

An important element of policy overseas concerns attitudes towards the use of chemicals (fertilizers, herbicides, insecticides). The use of herbicides and all but a few insecticides is increasingly banned either explicitly or implicitly. Thus in Denmark (Hansen, personal communication) and Sweden (Dahlin, personal communication) aerial application of herbicides is completely banned, while in Sweden herbicide control of weeds in coniferous regeneration is usually not permitted. The fact that herbicide use has become an issue in France (as noted by Martres, 1991) indicates how important it is: precautionary work to find the best practicable alternatives is clearly also desirable.

Up to 1991, genetically manipulated organisms, such as trees and insect predators or parasites, had not been used in the field for forestry purposes. The first introductions will call for the greatest care if the enormous potential of such developments is to be realized.

A desire, noted in Norway, to see more small machines used may be an appropriate strand of policy to develop in relation to small properties where the productivities of large machines cannot be exploited, and the aim accords well with moves towards more environment-friendly operations. Sweden has an association carrying out research and producing its own journal on small-scale forest machinery. But in neither country could it be said that the subject has risen to the level where it constitutes a point of policy requiring particular attention. Indeed it appears perverse to favour public assistance with the use of less economic means of carrying out operations which are in any event short term in their appearance at any particular site.

The planting of farmland

This topic raises acutely the need for coordination with other government policies: apart from familiar environmental concerns associated with change in the landscape, effects on physical factors such as water yield and quality, and impacts on employment, the displacement of farming has some special features. First, farmers are usually treated by governments differently from other entrepreneurs or workers; consequently income and work opportunities to this class of worker have to be given particular attention. Second, in Britain, the lack of any tradition of woodland management among most farmers creates peculiar conditions for implementation of planting.

EC Regulation 1096/88 authorized the provision of aid to countries' approved schemes for the afforestation of arable land. Farmers have shown little enthusiasm for the resulting schemes. However, the very short currency of existing or recent arrangements provides too little experience on which to base a judgement on the value of afforestation relative to alternative methods of restraining the growth of agricultural output. In addition, the first schemes have in part been used to test European Commission strategy as much as farmers' responses. More fundamentally, that strategy cannot develop coherently until the goals of the Common Agricultural Policy have been redefined in the light of the GATT round and internal EC debate.

The scale of change in use of land for agriculture is in any event unlikely to be as large as that suggested by the naive calculation of 'surplus' output divided by average production per hectare. As discussed by Harvey (1991) and Swinbank (1992) also, product price reductions will not have this scale of effect, and there will be shifts of ownership and land use which will slow the removal of resources from agriculture. Setaside of cereal-growing land with payment as an essential ingredient (as opposed to price reduction without compensation) will also be costly to the Community's taxpayers. Koester (1989) shows that, on plausible assumptions as to premium paid to farmers, the export restitution per ton of wheat sold and the proportion of the setaside premium provided by the Community, countries lose by the programme. Substantial progress in this field is unlikely until the means of shifting land use and the scale of change desired have been resolved, subjects on which politicians are understandably reticent. If and when a policy for larger shifts of land use is promulgated, an important consideration in formulating policy will be the origin of the investors undertaking planting and the form of land tenure adopted where the planters are not the farmers themselves.

The role of the state

Although the economic importance of state enterprises has declined markedly in many western countries in recent decades, and the state as an economic agent is rapidly withering away in former communist states, the policy on government control of private forestry in the countries under review has shown singularly little change. This is partly because at the same time as major political changes have occurred, there have been ever more clamant calls for environmentally desirable management in all fields of rural land use. Some decreases in financial support for forestry indicate a changing attitude to former imperatives for intensified wood production but this is the limit of change. No European country and relatively few of the American States have allowed market forces free rein in private forestry. On the contrary more consultation is called for than ever in all countries, more guidelines are introduced and more complex goals are set. None of the countries reviewed appear willing to loosen their control.

Goal Setting

Quantitative goals for private forestry are rarely set in the countries reviewed, the closest approach being in countries such as Sweden with industries that constitute important components of the economy. Here the maintenance of wood supplies at mutually acceptable prices has been an intermittent problem. Rarely is the figure for allowable cut taken as the sole guide to government action in these circumstances, and resort has only occasionally been had to subsidization of harvesting. By contrast, attention has been paid to economic difficulties of long standing in some regions, such as the working of distant forests, stands in steep terrain, and first thinnings. Equally, goals for new planting in the private sector have rarely been established. Where a plan exists, as in Denmark, the role of private planting is only indicated, whereas in The Netherlands there is no specification on the part to be played by the private planter. This is in contrast to the argument in Britain that fixing a target for afforestation over a period, and the rate this implies, is helpful to planners, potential investors and the public at large.

The forestry aims of all the countries reviewed in the course of the study are clearly in process of change. While the underlying economic conditions, as determined by wood price, have generally worsened, thus limiting the ability of private growers to finance operations, the claims of environment, and particularly those of conservation *sensu lato* and of recreation, have increased. Here goals are of an essentially qualitative nature and the valuation of benefits and quantification of targets are uncommon steps in the development of policy. Although attention to environmental aspects of management is increasingly accepted by private landowners, the fact that the outputs are not marketable introduces difficulties. Apart from the obvious one for governments of the need for more public finance, acceptance of such conditional aid reduces the freedom of action of owners.

Conclusions on the Guidance Offered on Policy Issues and Goals

The following conclusions can be drawn.

1. Whether a country has an afforestation policy or not, the central issues remain remarkably similar among countries.
2. The desirability of changing management to make forests more environmentally friendly has been accepted by all governments.
3. The adjustments required from private growers to produce more environmental goods and to reduce environmental bads grow greater with time.
4. The kind and scale of adjustment has usually been decided administratively in contrast to previous adjustments concerning wood production goals which were largely market-driven.
5. It is in the interests of forest authority and private owners alike to obtain more objective assessments of values in order to improve the basis for the determination of goals and incentives to implement policy.
6. Apart from the case of recreation provision on public lands, an objective basis for the type of change desired is generally lacking though the means of determining better allocations of resources are available.
7. Greater dialogue between forest owners and their managers on the one hand, special interest groups as a second party and the public as a third is required if rational decisions on policy and goals are to be taken.

References

Alston, R.M. (1990) Integrating economics and ecology: a case of intellectual imperialism? Review essay. *Forestry and Conservation History* July, 42–143.

Anderson, D. (1991) *The Forestry Industry and the Greenhouse Effect*. Scottish Forestry Trust and Forestry Commission, Edinburgh.

Anderson, T. and Leal, D.R. (1990) *Free Market Environmentalism*. Westview Press, Boulder.

Anon. (1990a) [*Report of the Interministerial Group on Warming*.] Ministère de l'environnement et de la prévention des risques technologiques et naturels majeurs. 83pp. plus annexes. Paris.

Anon. (1990b) [Tax on CO_2; interim report.] Office Fédéral de l'environnement, des forêts et du paysage, *et al.* Bern, Typescript.

Arnold, J.E.M. (1991) The long term global demand and supply of wood. Paper No. 3. In: *Forestry Expansion: a Study of Technical, Economic and Ecological Factors*. Forestry Commission, Edinburgh.

Aronson, R. (1992) Whose world is it anyway? *New Scientist* 7 March, 52–53.

Asp, A. (1990) Future fibre deficit in Europe calls for a change in strategy. *Pulp and Paper Magazine* No. 5/6 October, 26–31.

Beatley, T. (1991) A set of ethical principles to guide land use policy. *Land Use Policy* 8(1), 3–8.

Benson, J.F. and Willis, K.G. (1992) Valuing informal recreation on the Forestry

Commission estate. *Forestry Commission Bulletin* 104. HMSO, London.

CEC (Council of the European Commission) (1991) *A Community Strategy to Limit Carbon Dioxide Emissions and to Improve Energy Efficiency*. Commission proposal 8918/91 Brussels.

CEPI (Confederation of the European paper industry) (1992) *Forest Resources and the Community Paper Industry*. Brussels.

Dayton, L. (1990) New life for old forest. *New Scientist* 8 October, 25–29.

Elers Koch, N. (1978) [Forest recreation in Denmark: Part I The use of the country's forests by the population.] *Det forstlige Førsogsvesen i Danmark* 35(3), 287–451.

Elers Koch, N. (1980) [Forest recreation in Denmark: Part II The use of the forests considered regionally. *Det forstlige Førsogsvesen i Danmark* 37(2), 75–383.

Ellefson, P.V. (1992) *Forest Resources Policy*. McGraw-Hill, New York.

Forestry Commission (1984) *Broadleaved Woodland Guidelines*. Forestry Commission, Edinburgh.

Forestry Commission (1991) *Forests and Water Guidelines*. Forestry Commission, Edinburgh.

Franklin, J. (1989) Toward a new forestry. *American Forests* 95(11,12), 37–44.

Grayson, A.J. (1991) Review of IPCC Reports of Working Groups I, II and III. *Forestry* 64(4), 416–417.

Gregg, N.T. (1991) Will 'new forestry' save old forests? *American Forests* 97(9,10), 49–53, 70.

Grove-White, R. (1991) Politics, databases and vegetation science. Paper to British Ecological Society conference on The future of vegetation science. Typescript. Lancaster University.

Häckner, J. (1988) [The future competitiveness of bioenergy.] Dissertation, Stockholm School of Economics.

Harvey, D. (1991) The agricultural demand for land: its availability and cost for forestry. Paper No. 11 In: *Forestry Expansion: a Study of Technical, Economic and Ecological Factors*. Forestry Commission, Edinburgh.

Hultkrantz, L. (1989) The consequences for the forest sector of nuclear power discontinuation in Sweden. *Scandinavian Journal of Forest Research* 4, 395–406.

Hultkrantz, L. and Wibe, S. (1991) [*Forestry Policy for a New Century*.] Report to the expert group for studies in the public sector, Finance Ministry. Allmänna Förlaget, Stockholm.

Hummel, F.C. (ed.) (1984) *Forestry Policy: a Contribution to Resource Development*. Martinus Nijhoff, The Hague.

Interdepartementale Klimagruppen (1991) [*Greenhouse Effects and Responses*.] Miljødepartementet, Oslo.

Koester, U. (1989) Financial implications of the EC set-aside programme. *Journal of Agricultural Economics* 40(2), 240–248.

Lee, T. (1993) *Forests, Woodland and People's Preferences*. Forestry Commission, Edinburgh (in press).

Martres, J.L. (1991) [*Report*.] Syndicat des sylviculteurs du sud-ouest. Bordeaux.

Mattson, L. (1989) Hunting in Sweden: an empirical study of extent, motives, economic values and structural problems. Swedish Agricultural University, *Department of Forest Economics Arbetsrapport* 95, Umeå.

Nature Conservancy Council (1989) *Guidelines for Selection of Biological SSSIs*

(Sites of special scientific interest). NCC, Peterborough.

Navrud, S. (1989) Estimating social benefits of environmental improvements from reduced acid depositions: a contingent valuation survey. In: Folmer, H. and van Ieerland, E. (eds), *Valuation Methods and Policy Making in Environmental Economics*. Studies in Environmental Science 36. Elsevier, Amsterdam.

Norton, B.G. (1991) Review of three books on environmental ethics. *Journal of Forest and Conservation History* January, 39–40.

Opheim, T. (1991) The demand on research in forest economics from the policy maker's point of view. Paper to the Scandinavian Society of Forest Economics, Gausdal. Typescript. Norway.

Palmer, K.L. and Krupnick, A.J. (1991) Environmental costing and electric utilities' planning and investment. *Resources for the Future* no. 105. Washington, DC.

Pearce, D. (1991) Assessing the returns to the economy and to society from investments in forestry. Paper No. 14. In: *Forestry Expansion : a Study of Technical, Economic and Ecological Factors*. Forestry Commission, Edinburgh.

Pearce, D., Markyanda, A. and Barbier, E.B. (1989) *Blueprint for a Green Economy*. Earthscan Publications, London.

Price, C. (1978) *Landscape Economics*. Macmillan, London.

Solberg, B. *et al.* (1991) [*Forest and Forestry Production in Norway as a Counter to rising CO_2 Concentration*.] Report, Department of Forest Economics, Agricultural University, Ås.

Swinbank, A. (1992) A surplus of farmland? *Land Use Policy*, January, 3–7.

Usher, M.B. (ed.) (1986) *Wildlife Conservation Evaluation*. Chapman and Hall, London.

Usher, M.B., Brown, A.C. and Bedford, S.E. (1992) Plant species richness in farm woodlands. *Forestry* 65(1), 1–14.

Westoby, J. (1989) *Introduction to World Forestry*. Basil Blackwell, Oxford.

Willis, K.G. and Benson, J.F. (1989) Recreational values of forests. *Forestry* 62(2), 93–110.

Willis, K.G. and Garrod, G.D. (1991) An individual travel-cost method of evaluating forest recreation. *Journal of Agricultural Economics* 42(10), 33–42.

Winpenny, J.T. (1991) *Values for the Environment: a Guide to Economic Appraisal*. HMSO, London.

17

CONCLUSIONS ON POLICY IMPLEMENTATION

INTRODUCTION

A second theme of this study concerns the means and levels of financial incentives that certain member States of the European Community, together with other countries with similar economic and social backgrounds, adopt in order to encourage afforestation. This chapter reviews the literature on the rationale and effects of different types of support available to private owners through tax, grant, research and extension systems, and summarizes the financial aid available as assessed in the nine countries surveyed.

Inter-country comparisons are of interest in relation to the evolution of European Community practices. Already in the field of setaside and the afforestation of such land, the rates of direct grant paid by governments are, for obvious reasons, strongly influenced by Community policy on agricultural support. In addition although moves in the European Community towards harmonization of direct taxation are likely on present indications to be long delayed, convergence of tax systems, bases of assessment, etc. is clearly more likely to make later adjustment easier than the opposite.

The methods adopted to tax and aid investors are closely tied to existing systems of taxation and attitudes to grant aiding forestry and agriculture in the countries concerned. Since afforestation has been the exception rather than the rule in most western European countries over the post-war period, systems of financial support in these countries are usually geared to an established set of owners and forest assets. It might be thought that these circumstances must limit the relevance of any review of different countries' approaches. In practice, a remarkable similarity of approaches and tools is found among countries with very different agrarian histories, land tenure systems, population densities, forest estates and forestry policies. It is in the levels of taxes and more especially direct aids that differences are most obvious.

Policy Analysis

A small industry grew up in the 1970s concerned with the field of enquiry called policy analysis. In relation to forestry, the function of this discipline is to assess how successful government measures designed with particular forestry purposes in mind have been in achieving their aims. The term 'policy analysis' was originally adopted to refer to the study of all aspects of policy, that is from aims through to monitoring and analysing the results of public interventions. Although the term 'programme analysis' has been coined to cover studies of implementation, the more familiar term 'policy analysis' is employed here. This usage reflects the fact that the scope for analysis and the volume of work done is much greater in relation to policy implementation than to policy formulation.

In the present state of the art, which has been developed mainly by economists, there is no special capacity for any deep psychological understanding of the activities of forest owners in response to government policy although Kurtz and Lewis (1981), as noted earlier, showed one route to such an appraisal. Most enquiries use the empirical approach of attempting to discover relationships between government schemes and the responses of owners. This is no different from demand analysis and other techniques of market research applied to the study of consumer goods. The implementation of policy is an intensely behavioural matter but this does not mean that behaviour is unstable and unpredictable.

The findings of policy analytic studies may assist in devising the means of implementing new policy or improving the performance of an existing one. Publication of findings is fully in keeping with principles of open government including the notion that a democratic society should be able to discuss such matters as new initiatives of the administration rather than be faced with *faits accomplis* such as the dramatic change of policy on income taxation of forestry in Britain in 1988.

Ambitious claims about the ability of policy analysis to help decision makers in government have been made (see, for example, Järveläinen, 1981; Tikkanen, 1981, 1986; Ellefson, 1989) though others have been more modest in their claims (Riihinen, 1981). Since many government initiatives relate to the private sector, it is not surprising to find that most studies have focused on schemes aimed at results such as increased afforestation, adequate restocking and increased cutting. A useful stimulus to work in the field was given by the International Union of Forest Research Organizations in 1970 when a special working group (S.4.06-01, since 1991 known as S6.12-01 and entitled 'Analysis and evaluation of forestry policies and programs') was formed to consider the effectiveness of forest policy measures as applied to small woodlands. Regrettably only a few researchers have undertaken much work in this field, with Nordic and US workers prominent.

In many countries, among which Britain is increasingly one, interest centres on the questions of why owners cut the amounts of wood they do and how these commonly fall short of industry's capacity. Studies on cutting activities have been particularly common in Scandinavia because of the high government interest in the strategic concept linking more intensive wood production with the income and especially foreign-exchange earning ability of the wood-using industry. So far, those analysing the subject have only derived two clear findings. The first is that there is generally a reasonably firm relationship between price in any given year and the amount of wood sold, the second that old owners usually cut less than younger ones (see, for example, Lönnstedt 1986).

A second major field of concern has been that of restocking after cutting. This work links with that on afforestation and forest improvement, in which area a useful overview of US studies has been compiled by Alig *et al.* (1990). Because of the long delay in the production of returns from planting, these studies necessarily introduce questions of assumptions about future costs and prices and the hoary matter of choice of discount rate. This subject is returned to below under 'Grants'.

Ellefson and Wheatcraft (1983) have pointed out a deficiency of past studies in that there may be other goals than economic efficiency, such as equity, that motivate policy makers and thus sustain government support. Few studies have analysed how the results have applied to different regions and to different categories of owners, but Brooks (1986) provided an example of one such enquiry by assessing how increased wood production from grant-aided plantations altered the distribution of benefits between growers and consumers of roundwood. The rarity of this sort of study emphasizes once more the pervasive lack of enquiry on the basis of any given policy.

Another field of government intervention is that aimed at producing defined environmental effects which are broadly consistent with timber production aspects in planting and tending schemes. Thus far, little practical experience has been built up largely for the usual reason that hampers much environmental research, namely the absence of any adequate mensuration technique for many outputs.

Overall, the study of the assessment of effects of policy instruments, which is so central to the interface between government and private foresters, has been the subject of very little published research in European Community member States. The European Commission itself had the foresight to commission two studies, one (1976) on forestry matters including aids and taxes, the other (Guillard, 1978) dealing with small private woodlands. The lack of studies represents an unsatisfactory state of affairs. The disadvantages of becoming ensnared in a particular type of relationship or support system are too readily apparent from the warning provided by the Common Agricultural Policy for uncritical acceptance of the *status quo* to be sensible.

Direct Taxation

Indirect taxation on producers in European countries is increasingly covered by value added tax. As noted in Chapter 4, this is important in relation to consumers of forest products. So far as forest owners are concerned, the usual situation is that tax on inputs is recoverable. Discussion here therefore centres on direct taxation.

The forestry literature on intercountry comparisons of taxes is remarkably scanty. The material falls into two classes. One records the broad arrangements, the second analyses the supposed effects, of the prevailing tax system. In the first category, Kroth's (1963) study described the tax systems in seven European countries. In his introduction to the work, Speer noted that the Common Market aimed to harmonize economic conditions and asserted that 'The taxes and tax systems of the countries involved are a major obstacle to the achievement of that aim. No less important is the problem of the relative tax burden of forestry in the countries of the EEC, compared with those of EFTA and Finland.' Both statements must be regarded as highly questionable so far as direct tax differences are concerned. That different systems of direct taxation can distort competition is undeniable, but the idea that this has any influence on wood growing is more dubious. Similar thinking may have been the source of an earlier and in one way more ambitious study by Stridsberg and Algvere (1967) on eight west European countries. It aimed to set out a direct intercountry comparison of costs and revenues, based on accounting records and calculated for hypothetical models, but determination of the tax burden was found to be impracticable.

Hart's (1979) study represents a further important benchmark for European countries. The coverage was vast, including indirect taxes, but nevertheless the author was clearly restricted by the quantity and quality of the material placed at his disposal by the authorities. Although dated, the work represents an essential reference point and, in particular, illustrates clearly the importance of the detailed administrative rules that apply. No attempt was made to calculate the tax burden on forest owners in the 30 countries, eight of which had centrally planned economies, but instead a certain number of observations were made on whether the effects of the tax arrangements were considered favourable to increased wood supply or not. Harou (1990) provides a summary of forestry tax arrangements in force in the European Community and the US in more recent years. He concludes that income tax is the most important tax in the EC.

In his study forming part of a European Community review of current forestry problems, Kroth (1976) carried out an assessment of the hypothetical tax burden on existing forestry properties in Member States. Tax burdens were calculated for different sizes of woodland and for typical broadleaved and coniferous crops respectively under the prevailing tax laws, first, assuming the same costs and revenues in all countries, and second, assuming the actual costs and revenues applying in the particular country. No attempt was made to consider the tax aspects of afforestation. The results were clearly very strongly influenced by the cash flows assumed in the second set as well as by the tax rates applying in each country. The explicit assumption made in the study's recommendations was that the conservation and improvement of forests was to be achieved without any planned alteration in the ownership structure. This appears a generally acceptable assumption under the then and now current liberal economic concepts of private property ownership. Among the strong recommendations made was the desirability of avoidance of market values as the basis of assessment to capital and property taxes. This subject of the relevance of market values of woodlands for various measures is returned to later in the chapter.

Similar calculations are made by those engaged in government and in private forestry in connection with the consideration of amendments to existing taxes. Thus in Britain, following the introduction of capital transfer tax in 1975, the Forestry Commission made two sets of model calculations, one for a hectare of new planting, the other for existing woodland estates with normal and more or less skewed age-class distributions. These investigated the present value at a range of discount rates of differing total burdens and incidences over time of alternative methods of payment. Similar calculations were made preliminary to the introduction of increased grant rates following the removal of income tax and therefore the tax incentive to forestry in Britain in 1988. Spilsbury and Crockford (1989), writing after the change, indicate the results of such computations for a range of silvicultural systems, price levels, etc. Holten-Andersen (1985) provides a clear exposé of the Danish system of income and wealth taxes using a graphical presentation.

All such computations can assume unchanged forestry practices or ones which are influenced by the tax itself. Gamponia and Mendelsohn (1987) refer to this point, noting that if a tax does not alter management decisions then the burden it imposes is the present value of the revenue itself. They draw attention to the idea of 'excess burden', being the present value of lost income caused by distortions induced by the tax. Clearly it is reasonable to favour taxation which minimizes excess burden insofar as the outcome of owners' actions is in line with what society intends. Two tax arrangements were compared, one a tax on yield from cutting, the other an annual property tax. Their conclusion was that the impact on rotation length, the variable used to judge changed practice, under either of these was minimal.

The Gamponia and Mendelsohn (G-M) paper is an example of the genre which considers effects in order to draw normative conclusions. Many forest economists have found this a rewarding area; the impact of such research on policy is harder to assess. Thus, Klemperer (1987), responding to the G-M study, notes that a property tax which of its nature biasses against capital intensive land use distorts among land uses. Hiley (1930) observed that in the United States property tax was the chief deterrent to forestry, since it was a direct inducement to rapid felling, leading loggers to log an area fast and then lose their title to the land and so avoid any tax charge. Klemperer (1989), in his review of the investigations into forest taxes of the United States, points out that there were those who believed that property tax was shifted, in this case into stumpage values. But, given the degree of competition both nationally and worldwide, property taxes were more likely to be shifted back into lower land values (Klemperer, 1977). Amacher *et al.* (1991) illustrate what may be termed an economics engineering approach. The authors show how an investor bent on maximizing net present value over a whole crop life would adjust spacing and rotation length depending on the basis of taxation, the tax rate and the rate of interest. In a more realistic setting, Eid (1989) discusses the cutting decision in relation to the after tax rate of return and observes that because of the very low alternative rate of return in Norway there is little incentive to carry out final fellings, this being a major cause of the large build-up in Norwegian growing stock. Baardsen (1991), following the tradition of forest taxation studies at the Forestry Department of the Agricultural University of Norway, offers a prediction on the impact of the new Norwegian tax system introduced from January 1992 on cutting practice by private owners, suggesting that the cut will increase.

Klemperer (1989) observes, rather optimistically, that those who make forest tax policy

> need to know not only the effects of tax alternatives upon forestry but also the aggregate economy-wide effects. . . . Researchers need to examine and quantify net effects in order to know whether society as a whole gains or loses from a given tax policy. Research that considers forestry investments alone or that pursues taxes which promote forestry will tell only half the story.

Klemperer's own review gives a full statement of the work, almost wholly relating to the United States, on the inferred rationales for various taxes and their calculated, or supposed, effects. But the number of studies dealing with empirical investigations in the manner described below for various forms of grant aid in forestry is very small. A revealing exercise was that mounted by Hansson *et al.* (1988) as part of the project under the direction of Lönnstedt at Skog–Industri–Marknad Studier (or SIMS: forest–industry–market studies) Uppsala, concerned with private forest owners' decision-making systems. Five Swedish forest owners were asked to design their own 10-year working plan for a specific forest area, taking account of their own objectives and economic circumstances, including their personal tax position. Not one of the resulting plans corresponded to that favoured by the county forestry board. Despite the highly progressive Swedish tax system and very different volumes of cut, the marginal rate of income tax paid was not so different among the owners, presumably because of the variation in the other sources of income of the five.

The lack of such quasi-empirical studies on tax effects must stand as a criticism of research into this field and of the unqualified assertions that litter it. Such assertions are to be expected from interested parties but the lack of effort devoted to objective appraisal is unfortunate. However it would be idle to pretend that this is not a difficult field for policy analysis. Tax changes are not so frequent as some other alterations in the economic environment such as wood prices or interest rates and the scope for detecting change caused by one factor such as tax is therefore much reduced. Secondly, even where change has been systematic, as in the declining trend in the rate of income tax in most western countries over the past decade or more, the general econometric problem of identification of change with one particular factor is persistent. Thirdly, if a cross-sectional approach is adopted using the individual forest owner as the basis of variation, so that there is no difficulty about the number of observations, this problem of identifying the effect of a tax with only one factor such as income, rather than with income and other potentially relevant personal factors such as age, previous knowledge of forestry or size of holding, becomes acute.

Despite the difficulties, some useful findings have emerged from studies of the effects of particular tax regimes. Examples concerning cutting regimes are cited in Appendix III.

The inevitable conclusion of this short survey is that there is a dearth of empirical work on forestry taxation in Europe apart from that noted above being undertaken by Nordic workers. Even the coherent description of taxes in published form is lacking in some countries. If taxation is as important as many commentators in both industry and academe agree, it is astonishing that so little published work exists on the subject. Forest economists are often so heavily engaged in the justification of state intervention in forestry in one form or another that they have paid little attention to considering how well existing tax arrangements are working and what their effects are.

GRANTS

The term 'grant' is used in preference to 'subsidy'. The word subsidy is defined in national income accounts as relating to payments influencing current production. The European Commission has drawn attention to the practical importance of the distinction between subsidies, which affect the costs of current production, and longer-term expenditure, which does not, in connection with its concern over unfair competition through government support of R and D and launch costs in manufacturing industry. This consideration is relevant to the present discussion because of the occurrence among the payments noted in country chapters of subsidies to the transport of broadleaved roundwood and the extraction of conifers in the mountains of Norway, and on first thinnings in, for example, Denmark and Wallonia.

No distinction is drawn here between loans and outright grants. Among the countries surveyed, only France makes a major disbursement by way of loans and these are on such generous terms as to be tantamount to grants.

The literature on policy analysis in the field of grant aid is far richer than that on the appraisal of tax and policy influences on cut in the short term, that is from existing growing stock. The bulk of the work depends on quantitative material describing the actual behaviour of forest owners reacting to the combination of government regimes and their own circumstances. Regression techniques of one form or another are commonly used in contrast to the questionnaire approach which is limited in its ability to make any quantitative assessment of responses to grants. Valuable reviews of the American effort in this area are provided by De Steiguer (1983) and by De Steiguer and Royer (1986). The overall position reached from the latter survey in which some six papers on reforestation are reviewed is that good progress has been achieved in explaining the determinants of owners' behaviour. Tikkanen (1981, 1983), in the same field, indicates significant influences of grants, prospective income and owners' financial characteristics. The study by Kula (1988) on planting in Northern Ireland is also in harmony with others' research suggesting that planting grant levels have a major influence on planting rate. He computed a regression relating planting to a number of factors using a distributed lag procedure to take into account the fact that the rate of planting in a given year was influenced by the levels of some variables up to 5 or 6 years in the past and inferred an elasticity of planting with respect to real grant rate of 1.4.

These studies pay attention to the effectiveness of aids in achieving the immediate goals of policy such as more hectares of stocked forest. But others carry the argument the further stage of enquiring how valuable to society are the expenditures on public contributions of one kind or another; this is a much deeper concern for it is concerned with the merits of state support to the activity in question and not simply their immediate effect. Risbrudt *et al.* (1983) made an assessment covering private forest owners throughout the United States. They calculated the overall internal rates of return firstly to the direct costs of the planting or regeneration conducted, that is the owner's and the government's contribution, secondly to these costs plus overheads and thirdly to the owner's input alone. Their analysis suggesting that the scheme was effective did not pursue the point that large uptakes of government programmes cause big increases in future wood supply and hence price falls compared with the position where no government-induced planting takes place.

A critical contribution to the debate is that by Boyd *et al.* (1988) who, assessing the same programme, came to a very different conclusion. They calculated that in North Carolina, the subject of their exercise, the Forestry Incentives Program had spent \$1.618m of which only \$1.099m had been used effectively. Transfers to civil servants administering the schemes amounted to \$0.147m. By assessing the shift in the supply curve of stumpage when stands created as a result of the Program are cut and the resulting fall in price, it was possible to show that at a 4% discount rate[1], the gains to consumers would be of the order of \$0.372m and to growers \$0.301m. As with any public programme there is the so-called deadweight loss caused by the amount of subsidy that cannot be recaptured by benefits to consumers: this amounted to \$0.003m. Thus the overall loss to society equals 1.618 − (0.372 + 0.301) + 0.003 = \$0.948m. In this case one can deduce very crudely that, even if all the money had been used to regenerate and improve stands, the total benefits would not have amounted to more than about \$1.0m so that at best the programme would still have remained economically inefficient. Indeed, one might reasonably infer that if the provision of incentives shifting supply was so uneconomic, the current level of activity must be socially unjustifiable and ought to be taxed. The writers do not pursue this possibility, but note in addition, that to the extent that non-industrial private owners (NIPF) in the State have higher than average incomes, income distribution effects of the programme were undesirable.

This topic of the economic worthwhileness of different programmes of assistance is germane to the issue of the expansion of forests in the European Community, and especially that which is likely to occur on better quality land as a result of setaside. Existing published economic assessments, such as those of Hansen and Kjeldahl (1991) and Plochmann and Thoroe (1991) are essentially partial. Although Hansen and Kjeldahl very properly consider the possible influence on agricultural prices of GATT negotiations, neither study considers the possible impacts on wood price of large afforestation programmes in Europe. Both also ignore the environmental impacts of planting schemes. The ability of workers in the United States to assess efficiency depends on the fact that many of their programmes are intended to increase timber supply and the economic magnitudes relevant to estimation of wood price effects are broadly known. Where knowledge of the various demand and price elasticities is poor, or where the programme is concerned with environmental effects in addition to wood production, the ability to calculate social gains or losses of programmes is clearly reduced.

There are obvious scale economies in the delivery of aid and often also in its use. Targeting by size of forest has been determined in favour of the larger in France, whereas in Denmark farmers with small areas of woodland continue to be favoured. Few countries adopt the woodland size

---

[1] The discount rate is fairly critical, that adopted being the real rate used as the standard discount rate in US Forest Service appraisals. If r_1 is the rate of return expected pre-tax and pre-grant and r_2 the post-tax rate, then the rate of return on public funds must lie below r_1 and may be negative, as in Britain. The economic justification for support must be that r_1, and possibly r_2, understate the social rate of return.

differentiation found in British schemes, partly because a percentage basis of grant support is used which obviates much of the difficulty.

There is a battery of administrative questions to be considered relating to the effectiveness of alternative methods of delivering aid, namely whether through reduced taxes, grants, loans, gifts of material, provision of advice or other forms of cost sharing. In relation to grants in particular, the bulk of systems of payment used in Europe rely on stated percentages of costs, these being checked and agreed by the inspecting forester before operations begin. This method of payment, which is the norm in agriculture in Britain, is by contrast not favoured by forest authorities in Ireland and Britain. Second, it is obvious that the costs of administering grant aid differ between countries: the interesting question is whether some methods of organizing incentive programmes are cheaper than others. No studies have been found on these technical but nevertheless potentially important points.

Extension

The difficulty of finding statistics for public expenditure on extension in the countries surveyed may in part reflect an administrative difficulty, or may in part stem from a general lack of enthusiasm for study and control of the activity. Indeed the subject of the method and volume of extension work that is desirable is hardly aired among British foresters. This is in stark contrast to the tradition in agriculture where the farm adviser from the Ministry has become a familiar part of the countryside scene. Yet despite the value of his services and the long-standing interest of rural sociologists in the process of diffusion of agricultural techniques, the subject has not proved the most popular among students of agricultural economics. In the journals of the Agricultural Economics Society, Jones (1963) provided a benchmark contribution on diffusion and only in 1990 did a further review (Fearne, 1990) concentrating on communications, appear. One would have to look even harder for similar discussions in the British forestry literature.

The Forestry Act 1967 makes clear that, among the functions of the Forestry Commission, the promotion of forestry was to be regarded as being of high priority. In Germany, extension is an integral part of the job of a Land forest service forester whose services are provided free. In Norway, county and municipality board foresters are in a similar position. In Sweden, county forestry board staff make a clear distinction between four different target audiences: foresters, owners, manual workers and the general public; though the last category has to pay for the privilege. In France the CRPF are centrally concerned with extension. And in the US there is a major cooperative programme with participation and organization at national, state and local level which encompasses forestry along with other rural industries (Ellefson, 1992).

Programme analysis has not entirely ignored the subject of the effectiveness of extension efforts. Appendix III focuses attention on the subject of policy analysis with special reference to the role of extension as one input to the decision-making process of forest owners. The general conclusion from this survey of the literature is that there is some positive influence, but by no means a strong one, on the impact of extension. If, however, the cost of the service is low, its encouragement may be a worthwhile use of public funds. Extension is a field where a large contribution is made by private consultants. The question arises whether public intervention is necessary. Cubbage *et al.* (1993) observe that the market is segmented, with State foresters dealing with the smaller owners and consultants of various kinds with the larger properties. The practical outcome appears to satisfy equity considerations even if the efficiency argument is not resolved.

Another point of entry to the discussion is to consider the means adopted to assess the value of research programmes the final output of which is uptake of results by users. Following a thorough search of the US literature, Bengston (1985) reported that 'little work has been done on forestry innovations *per se*'. Although an excellent summary of actions required in the field of technology transfer was made by IUFRO group S6.08 (Moeller and Seal, 1984), no indication was made of the level of effort that is appropriate in this field.

This dearth of study is more strange in those countries which emphasize extension. Elsewhere, the position is as if it was considered that the forest owner was an educated man who could be relied on to read the technical literature and manage by himself; if he needed more advice he could buy it from a consultant. Assuming that the agricultural success story behind extension holds a lesson for foresters, it is right that quantitative assessment should be undertaken of the methods used in extension, their effectiveness as well as the desirable scale of this activity. Three types of approach are recommended. One is a case study approach, taking a particular innovation and by enquiry among forest managers and owners tracing its uptake, sources of information on the innovation, reasons for adoption, etc. A second is to interview owners in order to characterize their state of knowledge of forestry generally and to identify the various factors of age, education, etc. as well as use of consultants, advisory foresters, publications, attendance at meetings and courses. Thirdly, as part of wider studies of the effectiveness of different combinations of public inputs into private forestry, enquiries into owners' planting, thinning, felling, etc. activity should pay regard to their technical knowledge as one factor influencing choice. In addition to any findings on the overall success of extension, such studies would help to concentrate attention on the point of cost-effective provision of advice.

Research

The word 'research' in forestry is usually taken to refer to biological rather than economic or operational research. In practice research is also undertaken into economic aspects, including that of the public support of forestry, but very little on sociological aspects. It is assumed in what follows that all subjects of relevance to private forestry are considered in assessing the scale of research programmes.

Because this study is concerned with the public policy aspects of forestry, interest centres on the public expenditure component of research. As it happens, the bulk of research in the countries surveyed is financed from public sources.

Research evaluation is a subject which generated a substantial volume of study and literature in the 1980s. Where individual innovations are evaluated, it is almost certain that these are selected for study because of their high profitability. Whole programmes, unless they have a single, dominant objective, are rarely, perhaps never, evaluated. Even partial evaluations are uncommon. One assessment for British Forestry Commission research (Grayson, 1988) suggested a pattern of returns which appears characteristic. This is of a few very high return projects, a large mass of middling performance projects and some with low returns, with a few yielding a net negative result. Unless it is possible to identify the characteristics of the poor performers, for example that they all arise in one field of science or one scientist's area, it is difficult to see how management can be improved so as to include more of the potentially high achievers. But there is no reason to believe that the average rate of return on forestry research is all that high. Certainly there is no evidence to suggest that a switch of funds from research to other support measures might not be profitable. No recent publicly available surveys of research value were found in the countries visited.

Commentary on Public Financial Support in Review Countries

Overall, experience with policy analysis indicates that satisfactory tools for assessing the factors bearing on the uptake of public support do exist, and thus it is possible to deduce the extent to which different kinds and degrees of incentive have the most effect. A clear commitment to analysis is required if taxpayers' money is to be well spent. Monitoring is a necessary, but by no means sufficient, stage in this process. The earlier conclusion on the more advanced state of transatlantic knowledge of wood markets and the help this gives in assessing the actual merits of a programme aiming at increased wood production reinforces the need for continuing work on future timber trends. In addition, and as emphasized in Chapter 16, there continues to be a yawning gap in the evaluation of most forestry benefits apart from wood production. One concern must be with assessment of the social and cultural factors underlying the emergence of new demands, for instance the blossoming of demand for wildlife conservation relative to landscape. A second route which must be pursued is that of continuing efforts to make monetary evaluations of these non-market benefits.

There may, in addition, be circumstances where a cost-effectiveness measure would be adequate; the creation of recreational woodland of the style of the Netherlands Ringstad woodland scheme provides an example which may be applied to the new British community forests.

Summary of Support Measures in Nine European Countries

Consideration of grants or taxes alone is unfruitful, but equally it is impracticable without mounting a very large exercise to generate the numbers indicating their combined effects on the financial position and behaviour of typical owners or new investors in a given country. The principal concern of this study is, however, not with investment appraisal but with the identification of features of others' systems which are worthy of note.

Tax

The temptation to make intercountry comparisons of tax burdens per hectare must be resisted. There are many factors which render such comparisons meaningless, most notable among them the fact that there are wide variations in net income and capital values between countries. Even if tax is calculated as a percentage of net income, a realistic assessment has to take account of the departure of current income from the long-term average, especially marked in recent years owing to wood market conditions. What can be confirmed is that in the nine European countries surveyed, apart from Switzerland where income is particularly low and Ireland which does not charge income tax on forests, income tax represents the most important burden.

Reliefs abound in taxation systems and this is part of the reason for the relative importance of income tax: taxes on capital are greatly relieved. It is in this field of the forms of relief that interesting, and potentially important, questions lie. It should be noted that it is impracticable to measure the value of reliefs because the benchmark for a non-discriminatory regime is necessarily arbitrary; and any assumption about a different regime would demand a general equilibrium solution being found since values, particularly of assets, affect management and hence tax receipts.

Five reliefs are important in the nine countries, excluding Britain, for which forestry taxation has been described:

1. Standard or, in Switzerland, quasi-standard, values are used to assess income in all except Denmark, Norway and Sweden, and in the latter countries various methods of deductions lower the resulting charge; standard valuations, like the former property values in Britain, are updated only intermittently and in Germany rarely.
2. Over and above the use of standard values, France is notable in providing relief for young woods.
3. Forests are relieved of property taxes in three countries and standard values are adopted as the basis for assessment in the remainder.
4. As of 1991, only two countries (Belgium and Ireland) had no wealth tax; in Denmark 80% relief on current market value is the base, in Germany standard value is used, and in the remainder notional capital or standard values were variously relieved. The Netherlands are remarkable in providing relief for a particular non-wood, non-revenue generating end, that of public access.
5. All the countries surveyed, except for certain Swiss cantons, adopt an inheritance basis for transfer tax rather than an estate tax; reliefs are broadly as for wealth with, however, various differences related to the fact that a different owner is involved.

The use of 'standard' values obscures the fact that the system has administrative inconveniences, one being the heavy task of revision, and introduces fiscal anomalies. Countries have the ready alternative of using market returns or capital values and providing relief on these values. This arrangement avoids the complications and arbitrarinesses of standard values which are elaborated only periodically and are liable to show unrealistic relativities even between properties in the same district as well as in the interval before the next revision.

Grants

Table 17.1 shows private forest areas and the costs of public expenditure in the nine countries surveyed and Britain. It should be emphasized that the figures relate to a single year, and that arbitrary attributions to private woodlands have had to be made. The fact that in one, Ireland, a large afforestation programme is undertaken relative to the existing forest is not significant except in one regard, namely that the rate has risen greatly in recent years and accordingly the balance of different forms of assistance cannot be considered as stable as in a country such as Germany.

The scale of grant payments cannot be realistically compared between countries which have markedly different programme emphases, for example with respect to afforestation. Leaving aside Ireland and Britain for this reason, inspection shows that the payments per hectare differ very widely.

Table 17.1. Public expenditure support of forestry in ten countries.

Country	Area of private forest (m ha)	Public expenditure total attributable to private forests[a]				
		Grants (m ECU)	Research (m ECU)	Ratio to grants	Extension (m ECU)	Ratio to grants
Belgium	0.33	0.41	2.5	6.0	N.av.	
Denmark	0.33	4.7	2.5	0.5	2.0	0.4
France	10.2	22	18	0.8	13	0.6
Germany	3.33	77	7	0.1	N.av.	
Ireland	0.095	14	1.1	0.08	0.7	0.05
Netherlands	0.136[b]	15	2.7	0.2	N.av.	
Norway	5.1	45	5.6	0.1	3.2	0.07
Sweden	17.2[b]	38	29	0.75	5.6	0.15
Switzerland	0.32	25	5	0.2	13	0.5
Britain	1.27	27	13	0.5	*ca.* 3	*ca.* 0.1

[a] In 1990 or 1991; conversion to ECU made at the rates listed in Appendix IV for January 1992.
[b] Excluding properties of nature protection societies.
[c] Including industrial private forests; grant payments falling rapidly.

The reasons for the relative positions of Sweden with 2 ECU ha^{-1} and The Netherlands with 100 ECU are not hard to discern, but the contrasts remain remarkable.

A second point of interest attaches to the ratios for costs of research attributed to private forestry relative to expenditure on grants. By the first of these measures, Britain is near the top of the scale with France and Sweden a little higher owing to the recent reduction of grant payments but no such diminution of research expenditure. Belgium stands alone because of the recency of the introduction of grants.

Less reliance can be placed on ratios measuring expenditure on extension relative to either grants or research owing to uncertainties of measurement of this variable. Nevertheless some interesting comparisons emerge. These once again emphasize the lack of understanding of the reasons for variations in these forms of expenditure among countries.

Conclusions on Financial Aspects of Implementation Measures

A full assessment would consider tax relief in cash terms and place this alongside direct public expenditure. It is not necessarily realistic to add the two since owners may, and commonly act as if they do, value £1 of tax relief more highly than £1 of direct post-tax aid. But there are more important reasons for considering the two forms of intervention as distinct. In the first place, grants are usually under the control of a single agency, the forest authority, which makes payments and exerts that authority in order to stimulate or depress activity in a certain direction. Good examples are provided by the Scandinavian emphasis on road construction, fertilization, drainage and planting in order to encourage wood production in the 1960s to 1980s, now moving towards a set of grants aimed at the provision of an environmentally friendly mix of activities. The same shift has occurred in Britain and indeed most of the countries surveyed, with the exception of France where the need to improve the cost-effectiveness of the Fonds Forestier franc to achieve wood production appears to have been almost the only adjustment of goals.

By contrast, special taxes or amendments to taxes constitute a far less attractive tool of policy for the forest authority to pray in aid. In the first place, provisions which favour one class of taxpayer are immediately the cause of complaint based on the notion of unfairness and the emotion of envy. Secondly, tax incentives are obviously very imperfect means of directing attention to particular forms of spending, in which specific grants are clearly advantageous. In addition the tax route suffers from the disadvantage that it is not transparent, a deficiency which makes monitoring of the efficiency of this route to the promotion of public policy difficult.

References

Alig, R.J., Lee, K.J. and Moulton, R.J. (1990) Likelihood of timber management on nonindustrial private forests: evidence from research studies. *USDA Forest Service General Technical Report* SE-60.

Amacher, G.S., Brazee, R.J. and Thomson, T.A. (1991) The effect of forest productivity taxes on timber stand investment and rotation length. *Forest Science* 37, 1099–1118.

Baardsen, S. (1991) [Maintenance of capital, with reference to tax reform.] Typescript, Ås. 6pp.

Bengston, D.F. (1985) Diffusion of innovations in forestry and forest products: review of the literature. In: Forestry research evaluation, *USDA Forest Service General Technical Report* NC 104, 62–68.

Boyd, R.G., Daniels, B.J., Fallon, R. and Hyde, W.F. (1988) Measuring the effectiveness of public forestry assistance programs. *Forest Ecology and Management* 23(4), 295–309.

Brooks, D.J. (1986) Evaluating the regional and distributional impacts of forestry cost-share payments. In: Analysis and evaluation of public forestry policies. *Silva Fennica* 20(4), 319–330.

CEC (1976) Forestry problems and their implications for the environment in the member States of the EC. Results and recommendations. *Information on Agriculture* no. 25, Luxembourg.

Cubbage, F.W., O'Laughlin, J. and Bullock, C.S. (1993) *Forest Resource Policy.* Wiley.

De Steiguer, J.E. (1983) The influence of incentive programs on nonindustrial investment. In: Royer, J.P. and Risbrudt, C.D. (eds), *Nonindustrial Private Forests.* Symposium Proceedings, School of Forestry and Environmental Studies, Duke University, Durham, North Carolina.

De Steiguer, J.E. and Royer, J.P. (1986) Increasing forestry investments by means of public policy programs. In: Analysis and evaluation of public forest policies. *Silva Fennica* 20(4), 354–357.

Eid, J. (1989) Maturity, nominal and real interest rate before and after tax. Typescript, Ås.

Ellefson, P.V. (1989) Preface in *Forestry Resource Economics and Policy Research.* Westview Press, Boulder.

Ellefson, P.V. (1992) *Forest Resources Policy: Process, Participants and Programs.* McGraw-Hill, New York.

Ellefson, P.V. and Wheatcraft, A.M. (1983) Equity and the allocation of Forestry Incentive Program funds. In: Royer, J.P. and Risbrudt C. (eds), *Nonindustrial Private Forests.* Symposium Proceedings, School of Forestry and Environmental Studies, Duke University, Durham, North Carolina.

Fearne, A.P. (1990) Communications in agriculture: results of a farmer survey. *Journal of Agricultural Economics* 41(3), 371–380.

Gamponia, V. and Mendelsohn, R. (1987) The economic efficiency of forest taxes. *Forest Science* 33(2), 367–378.

Grayson, A.J. (1988) Evaluation of research. *Forestry Commission Occasional Paper* 15. Forestry Commission, Edinburgh.

Guillard, J. (1978) [*The Role of the Small Private Forest.*] Étude P.E. 191. INRA, Nancy.

Hansen, J.K. and Kjeldahl, R. (1991) [Afforestation compared with agriculture.] *Statens Jordbrugsøkonomiske Institut Rapport* nr. 60. Copenhagen.

Hansson, K., Lönnstedt, L. and Svensson, J. (1988) Tax situation for private Swedish forest owners: economic effects of different forest management strategies. *SLU/SIMS Report* no. 2, Uppsala.

Harou, P.A. (1990) Forestry taxes and subsidies for integrated land use. In: Whitby, M.C. and Dawson, P.J. (eds) *Proceedings of the 20th Symposium of the European Association of Agricultural Economists.*

Hart, C. (1979) *Effect of Taxation on Forest Management and Roundwood Supply in Europe.* Report to joint ECE/FAO working party on forest economics and statistics, Geneva.

Hiley, W.E. (1930) *The Economics of Forestry.* Clarendon Press, Oxford.

Holten-Andersen, P. (1985) [The Danish income and wealth tax systems illustrated by three-dimensional graphs.] *Dansk Skovbrugs Tidsskrift* 70, 206–218.

Järveläinen, V.-P. (1981) Aspects of research strategy in studying forest owners' behaviour. In: Effectiveness of forest policy on small woodlands. *Silva Fennica* 15(1), 25–29.

Jones, G.E. (1963) The diffusion of agricultural innovations. *Journal of Agricultural Economics* 15(34), 387–404.

Klemperer, W.D. (1977) An economic analysis of the case against ad valorem property taxation in forestry: a comment. *National Tax Journal* 30(4), 469.

Klemperer, W.D. (1987) The economic efficiency of forest taxes: a comment. *Forest Science* 33(2), 379–380.

Klemperer, W.D. (1989) Taxation of forest products and forest resources. In: Ellefson, P.V. (ed.), *Forest Resource Economics and Policy Research*. Westview Press, Boulder, Colorado.

Kroth, W. (1963) *Systems of Forest Taxation and Tax Impact upon Private Forestry in Several European Countries*. Translation of the Israel Program for Scientific Translations, Jerusalem.

Kroth, W. (1976) Taxation. In: Forestry problems and their implications for the environment in the member States of the E.C. 1. Results and recommendations. *Information on Agriculture* no. 25, Luxembourg.

Kula, E. (1988) *The Economics of Forestry: Modern Theory and Practice*. Croom Helm, London.

Kurtz, W.B. and Lewis, B.J. (1981) Decision making framework for NIPF: an application in the Missouri Ozarks. *Journal of Forestry* 79(5), 285–288.

Lönnstedt, L. (1986) [Cutting decisions by nonindustrial private forest owners - an analysis of an interview investigation.] Swedish Agricultural University, *Institute for Forest Economics Report* 165, Uppsala.

Moeller, and Seal, D.T. (1984) Technology transfer in forestry. *Forestry Commission Bulletin* 61. HMSO, London.

Plochmann, R. and Thoroe, C. (1991) [Support of afforestation.] *Schriftenreihe des Bundesministers für Ernährung*, Reihe A, Heft 397. Landwirtschaftverlag, Münster.

Riihinen, P. (1981) Effectiveness of forest taxation reform as a means of economic policy. In: Effectiveness of forest policy on small woodlands. *Silva Fennica* 15(1), 92–99.

Risbrudt, C.D., Kaiser, H.F. and Ellefson, P.V. (1983) Cost effectiveness of the 1979 Forestry Incentives Program. *Journal of Forestry* 81(5), 298–303.

Spilsbury, M.J. and Crockford, K.J. (1989) Woodland economics and the 1988 budget. *Quarterly Journal of Forestry* 83(1), 25–32.

Stridsberg, E. and Algvere, K.V. (1967) Cost studies in European forestry. *Studia Forestalia Suecica* no. 49. Royal College of Forestry, Stockholm.

Tikkanen, I. (1981) Effects of public forest policy in Finland: an econometric approach to empirical policy analysis. In: Effectiveness of forest policy on small woodlands, *Silva Fennica* 15(1), 38–64.

Tikkanen, I. (1983) Effectiveness of forest policy programs: a policy analysis perspective in NIPF. In: Royer, J.P. and Risbrudt, C.D. (eds), *Nonindustrial Private Forests*. Symposium Proceedings, School of Forestry and Environmental Studies, Duke University, North Carolina.

Tikkanen, I. (1986) Search for innovative forest policies and programs: the future and role of policy and program analysis. In: Analysis and evaluation of public forest policies. *Silva Fennica* 20(4), 265–269.

18

Implications of Findings for the Framing and Implementation of Policy in Britain

Introduction

All west European countries surveyed in this study reject a free market solution to the management of private forests. Given that a national policy towards private forestry has to be established and adjusted from time to time, the challenge is to find better solutions to the questions that arise at all points in policy debates from determining aims to deciding the details of implementation. There are more than two parties to many of the debates that arise in connection with policy aims and clearly many more parties, both at national and at local level, involved in the implementation of policy. There are, however, only two principals who have the job of agreeing on objectives and methods of implementation: the government, represented by the forest authority, and the growers' organization. This chapter summarizes functions that serve the needs of developing, implementing and monitoring policy and which fall to be carried out by one or other of these bodies.

The Organization of Private Forestry Adminstration

In any government organization, politics inevitably have a major role to play. The greater weight given to *vox pop* in recent decades and the generally greater awareness of people about matters concerning their physical environment might be thought to have major implications for the way in which forestry is organized in government. To serve the wider constituency beyond tree growers themselves, the government might well wish to remove the forestry organization from a ministry which is tarred with a productivist brush, this despite the common association of food with agriculture in departments' titles. In fact, most forest authorities remain part of the Ministry of Agriculture.

Foresters' resistance to a move away from Agriculture has been strong in many countries, but the shifts to Environment which have occurred in Denmark, Flanders and The Netherlands (the last representing a halfway house in this connection) are of great interest and deserve to be monitored. One common objection to placing forestry within the ministry responsible for the environment is that such departments have usually had an urban rather than a rural bias. This criticism is valid to the extent that the bulk of professional staff have little concept of what is involved in rural estate management. Increasingly, however, the concerns of such departments are with the whole land and water base and not just the built environment. Despite the apparent attractions of such new dispositions, it is clear that in the countries with such an arrangement, the situation is hardly a happy one. Further, the relatively new activity of afforesting setaside or other non-marginal agricultural land is already proving a severe test of the theory that it is appropriate to place forestry within the environment ministry instead of agriculture. The administrative complications and resulting tensions for both foresters and farmers are much magnified where responsibilities are split.

The forest service of a country has traditionally had three functions:

1. Regulation as forest authority, covering services common to all ownerships such as the control of felling, the maintenance of environmental standards, research, etc.
2. Administration and supervision of private woodlands schemes.
3. Management of the state forestry enterprise.

With the increasing tendency to separate the enterprise and thus leave the forest service with the first and second functions only, questions are raised about the appropriate locations and staffing of the two activities. In theory there is no reason why the private woodlands branch, whether acting in a sponsoring or a controlling role, must lie within the authority. However, in only one of the countries considered here is there an example where the authority role is separate from that of private woodlands supervision. Partly this is because of the scale diseconomy of any such split. In Germany, where forestry is devolved to regional level through the Länder, private woodlands administration is sometimes hived off and the authority function then left with the forestry enterprise. The most persuasive argument in these matters concerns the scale of the forestry activity *in toto* and that of private forestry within it. If the scale of forestry is small, the opportunity for maintaining a cost-effective authority function separate from the private forestry service is remote. The alternative arrangement, that of placing the authority function within some larger division, possibly of a different department, is likely to be detrimental to forestry interests because the weight given to the forestry viewpoint is then small to the point of vanishing.

The establishment of a separate cadre of staff in a forest authority in Britain from 1 April 1992 (Forestry Commission, 1992) might appear to make the subject of its organization one of academic interest only. But there remain questions concerning staffing which deserve attention if the authority is to sustain high professional standards. For example, means must be found of ensuring that staff maintain their awareness of the economic realities of the forestry business, technical innovations and the like. The alternative of a stand-alone forest authority able to recruit its own personnel is clearly feasible in countries with a large non-state forestry sector supported by substantial economic interests, as in Norway and Sweden, but risky elsewhere for the same reason as noted above in relation to the hiving-off of a 'pure' forest authority.

The Process of Policy Review

Policy, whether in its aims or its mode of implementation, can never remain static. The following discussion concentrates on matters relating to form, while content is referred to in the penultimate section on policy analysis.

Methods of reviewing policy and proposing new policy aims are handled very differently in the countries studied. None of these ties itself to a regular review. In the course of their reviews of policy, countries may undertake exhaustive studies. For example, the assessments for Norway and Sweden in train in 1990–92 and those informing the debate which led to the new Forest and Nature Acts of 1989 in Denmark were far-reaching and included the active participation of a number of interest groups. It may be noted that in the USA the Forest Service is obliged by the Resources Planning Act of 1974 to undertake a comprehensive Assessment of the country's renewable resources every 10 years and to establish a recommended programme for the Forest Service every five. The Assessment requires the Forest Service to consider outputs of forest goods and services from all ownerships.

The methods of review adopted in several countries reviewed contrast with the British tradition as established until now. It is almost always advantageous, and may indeed be essential, to make the process of policy formulation open. In Britain the convention has been that ministers are best placed to judge what people want and thus give guidance to civil servants, who are then set to work to produce a policy consistent with that guidance[1]. This is

---

[1] The arguments for the chosen policy are rarely published although this course was partially adopted with the cost–benefit analysis of 1972 (HM Treasury, 1972). This document contains one of the few published official assessments of the costs of public expenditure support of private forestry, an assessment which in the outcome had no policy impact. There is no evidence that the important decision announced in 1988 to remove forestry from income tax was taken as the result of careful consideration: there was certainly no prior consultation.

hardly an acceptable way of doing business in areas of government which are not by any stretch of the imagination ones where sensitive matters of national interest are concerned. The Parliamentary Select Committee route is one method of opening up the debate and it may help in one function of policy review, namely investigating the state of forestry. More important, even when the timing is satisfactory, the purpose of the Select Committee system, which is principally concerned with assessing the working of policy in a particular area, constitutes only a small part of a policy review's proper function. In addition, the adversarial approach of such committees is hardly the best way of eliciting the most constructive or informed replies. The Australian, Dutch, New Zealand, Norwegian and Swedish systems all have elements which are desirable if a way is to be found of achieving a more widely based, participatory system of analysis, appraisal and debate as advocated by, among others, the Agriculture Committee (1989).

Mention has been made elsewhere of the unstructured way in which the agenda of reviews are usually determined. Politicians' current concerns are not the most desirable base although these are difficult to avoid as one starting point. Deeper consideration should be given to the use of disciplined appraisals of potential objectives and, where appropriate, their means of implementation. Open conferences are a practical means of gathering opinions on issues but further thinking is needed on the more difficult task of formulating national and, where appropriate, sector (ownership) objectives. Preceding chapters indicate that there are methods used in other countries which have useful features. Two constituencies apart from government are engaged: first the owners and secondly those with an interest in forestry arising from forestry's actual or perceived influence on forest products industry, employment, environment or other activity. Although consensus is clearly a most desirable way of progressing policy discussions, there are occasions when it is unrealistic to expect to achieve such harmony. At certain points, hard decisions have to be taken which may be unattractive to one or other party to the debate.

It is also important that those from the government side and special interest groups are fully aware of the reality of the owners' position. A continuous effort is needed to inform all parties, and not only at the highest level, about forestry issues and all means should be used to exchange facts and views with the public and single interest groups in particular. A case in point arose in 1988 when it became clear that government did not fully appreciate the role of income tax in assisting the financing of forestry operations unconnected with large-scale afforestation. To achieve the desirable understanding requires both a willingness to listen and the supply of information, including financial data, by the private sector.

The recent shift in practice represented by the issue (Forestry Commission, 1991a) of views of outside experts on the basic facts which underly policy and a subsequent conference in 1992 is a welcome departure. This represents a refreshing change of practice not seen in Britain since 1976 when an interdepartmental committee took evidence, though on a more restricted footing.

Apart from purely forestry subjects, policy reviews necessarily involve other fields in which government plays an important or dominant role such as land use, industry and taxation. Private forestry in Britain has become obsessed with taxation during recent decades, first with refinements of concessions under Schedule D of income tax, then with capital transfer tax and, later, inheritance tax, and finally with the aftermath of the abrupt removal of forestry from income tax. It is difficult to avoid the annual tinkering with inheritance tax that occurs, a process in which forestry has played a full part. If, as suggested below, there are sound arguments for reintroducing income tax on commercial woodlands, it is desirable that this should be openly discussed and considered by the suggested review committee. Tax initiatives have been publicly aired in the past before legislation has been introduced, or not, as the case might be. It is unusual to debate a tax regime for a particular industry in a way that is isolated from tax policy generally, but this represents a precedent only and hardly raises any fundamental constitutional matter. If taxation is a source of political difference, there is a risk that the policy review process can be blocked and in particular the participatory approach outlined rendered difficult. This does not imply however that the tax treatment of forestry should be handled as if it were a separate matter from other forestry issues; it is such a fundamental part of policy implementation that the issue cannot be left to one side.

Policy formulation cannot realistically be conducted without regard to international obligations: currently this means that European Community aspects must be considered. While there is a lack of strategic initiatives from this quarter, any such aspects will normally be confined to implementation.

Promulgation of the results of a more participatory form of review is easy. Except in the most unusual circumstances, and even with major swings of government, as occurred in Sweden one year before the last policy review, a desire to achieve agreement in the review process reduces the need for strong political debate.

The manner in which policy is stated differs enormously. As Hummel and van Maaren (1984) remark, a short statement is attractive but if it is too vague it is useless; if it is too precise it loses impact and may require frequent change. Unfortunately too many forest policy statements fail to set out the role expected to be played by different ownership groups. Forests can never be all things to all men. Any given forest can be likened to a musical instrument; some works can be performed well on a given instrument, some hopelessly badly. A plurality of owners, a range of sites and a spectrum of species and ages of tree combine to make it possible to produce a diversity of forests offering different satisfactions to their owners, as well as other beneficiaries. The limits of what one particular instrument, or type of woodland, can perform should be recognized and aims not set which are beyond the woodland owner's compass. Not all musical effects can be obtained if one removes a whole class of instruments, such as the woodwind, from the orchestra. So it is if one attempts to rely on one class of owner only, such as private individuals, rather than the state or local authority, to create or run a facility such as a community woodland. The view that private owners will contribute substantially to the creation and maintenance of such woodland represents a high water mark of belief in the ability of private enterprise to meet every challenge.

The failure to identify the place of the different agents within forestry is probably the biggest single deficiency of the statements recorded in the country chapters above.

Formulation of Goals

The term goal is used here to apply to the more specific description of objectives. Confidence within forestry is increased if owners and managers have a clear set of goals. Where practicable, goals will be expressed quantitatively, for example as the annual rate of afforestation proposed as the goal for the private sector, or the number of new recreational facilities or area of positive conservation areas such as new 'native' forests to be created by the state forest enterprise. More qualitative guides stand as goals where quantification is either inappropriate or currently impracticable; an example is the codification of operational standards.

The setting of goals by a broad-based review body, as opposed to a group of government officials responsible to ministers, is neither practicable nor desirable. The government of the day has the ultimate responsibility through the budget for determining public expenditure on state and private forestry. A review body, even one including members of parliament, can be helpful on emphasis but not decisive on the point of finance. There will, however, be some correlation between the attitude of government on the expenditure front and the will of the review body, and any committee that ignored the current political and economic realities would be failing in its task.

In Britain's circumstances a critical goal is that for afforestation. Past decisions on planting programmes have, however, lacked a logical basis. Although a substantial effort may have been devoted, if too little debated in national fora, to establishing a rationale for the desired scale of the Midlands forest in England and the number and sizes of community forests, surprisingly little such effort has been taken to develop a rationale for the programme of planting of new forest aimed principally at the production of wood. In Britain the Institute of Chartered Foresters (Agriculture Committee, 1989a) has been a strong advocate of the virtues of an agreed long-term programme of new planting and of one method of determining it. The argument used relies on the twin assumptions that it is possible to establish a sound reason for selecting some particular degree of self-sufficiency in domestic wood production and a particular date by which this is to be attained. Apart from the essential arbitrariness of the choice of the two figures implied, the proposition is liable to be circular, because any target that leads to annual planting rates well in excess of likely land availability lacks credibility. It is not a wholly valid criticism that other outputs than wood are ignored in the arithmetic, since many requirements in other fields can be subsumed within a planting programme driven by the wish to grow more wood. The fact that the argument does not extend to programmes for the creation of distinct types of woodland, such as community or native tree woodlands which are not driven by wood supply considerations, is more critical.

But the most significant gap in the argument as stated lies in the divorce from reality implied in ignoring its cost. There are two economic aspects to this. One relates to the resource costs of a given size and location of programme, the other to the financial implications for public expenditure. Further work aiming at a more cogent basis for a goal is essential along the lines of the economic appraisals presented in Dewar (1991) and Pearce (1991). The merit of the whole view to be taken by their approach is that recognition is given to all aspects of the gains achieved and losses incurred by continued afforestation. Explicit valuations of the benefits and costs concerned are not only used in relation to new planting but become incorporated into the management of existing woodlands[1]. The opportunity to achieve a coherent balancing of the claims of all interests is a stimulating one. Most thinking foresters agree that plantations of almost any species, but especially those used in first rotation crops which often produce dense shade and are sited on land with no tree or shrub seed bank, can and should be made more attractive for recreation and wildlife. Their desire for an agreed basis for their treatment, usually of diversification of species and edge, has been in part met by the issue of environmental guidelines, but these still do not satisfy either all forest managers or many outside commentators. The challenge to produce a more rigorous basis for deciding the composition and pattern of planting based on appropriate weights for all the outputs of the forest is a stimulating one. It is, arguably, more important than whether Britain plants 20,000 or 40,000 ha year^{-1}.

It follows from the above argument, which assumes intermittent assessment of policy objectives and costs of achieving them, that the concept that

[1] Such analyses must be the answer to those commentators (e.g. Mason, 1989) who favour a deregulated unsubsidized private market informing all decisions in forestry.

there is an optimum expansion path for forest area and that it is possible to set it is unrealistic. This is not to deny that at any one time it is valuable to have a desirable rate of new planting in mind for some years ahead since this allows rational planning by the different agencies such as nurserymen, investment advisers, contractors, regional or county authorities, etc.

The effect on wood, or, more important to woodland owners, revenue production that changed management of existing woodlands can make has been so low on the agenda of recent policy discussions as to be almost forgotten. As the stock of existing woods increases, so the impacts of decisions by their owners become more and more relevant to wood-using industry. Most of today's pressures, from landscape through carbon storage to recreational interest, imply postponement or reduction of the cut instead of the reverse. The impacts of reduced yield on the prospective supply of wood to processors and on revenue to owners must therefore form part of the considerations leading to a particular goal.

Technical factors may limit the possible contribution of woodlands to meeting certain aims. Thus the potential for creating a 'richer' forest is greater in countries with many species easily capable of being regenerated naturally. This propensity may be limited in much of Britain's new estate and it is necessary to inform the public of the physical and biological realities.

Implementation Measures

There are two elements in the implementation of policy. One is the set of rules for critical activities such as felling and restocking, the location of new planting or the use of particular species. Another is the system of incentives intended to encourage the desired scale and kind of activity. These include the tax system, other financial tools, principally grants, and services such as research information.

Controls

It is clear that a policy only requires to be established where the State has a role to play in controlling or encouraging an activity. This arises where there are market failures of one kind or another. The control aspect is philosophically the more difficult aspect of public engagement with or in an activity. In order to justify it, the case must be accepted that the free play of market forces will produce undesirable consequences. But how far interference should extend is a reasonable question, and one that is put more and more frequently today. Some aspects of national policy are accepted as applying equally to private owners and to the state forest enterprise. Restocking after felling is one example, and obedience to certain silvicultural and engineering rules for environmentally responsible behaviour another. There are in practice differing extents of the probing that is being done and of the pruning of control procedures that is being proposed. At one level the mothering attitude is discussed, as in Lönnstedt's (1991) critique of the restrictive cutting rules in Sweden and as in Switzerland on the matter of strict compensatory planting after forest clearance. In relation to restocking after felling, it may be asserted that if a different form of land use is proposed then the burden of proof in favour of the change must lie with the proposer. This was the view taken by Scott (1942) where the committee on planning and agricultural land use considered that primacy should be given to agriculture, a view resisted in a minority report by Dennison. The latter was not accepted at the time but, in the light of the current position of surplus food production, has recently come to be reconsidered. Although it is difficult to see how a policy aimed at increasing the flow of forestry goods and services can be plausible unless it requires that felling should be followed by restocking, the opening up of land use questions suggests that there is room for a less restrictive attitude. Where compensatory planting is feasible and permitted, a change in the disposition of woodlands may be desirable on landscape grounds.

No critique similar in depth to those in Sweden has been conducted on the British guidelines for forestry practices and water, landscape and nature conservation. As noted in Chapter 16, the precise standards set in the environmental field may well be arbitrary. Although the pressures for environmentally friendly measures are clearly increasing in all countries, cultural differences between countries and even regions of countries (as in France and Britain or Belgium and Britain) account for much variation in concern with environmental aspects of land use. The degree of control exerted over practices in countries such as The Netherlands in respect of the use of herbicides and Belgium, as of 1991, in relation to the range of species which may be planted on setaside land is remarkable. To the extent that these controls have been instituted through lack of political courage or through ignorance they are to be deplored. But the fault is not entirely with the politicians; the public belief in science has been shaken too often to ignore its worries. (Grove-White, 1991 goes so far as to suggest that unease over the impacts and future trajectory of industrial society has been the source of the environmental movements of recent decades.) There is no ready solution to this problem of the failure of minds to meet. If land users are to convince people that their practices are rational and safe, more contact is essential. Educational efforts through discussion, courses, woodland walks and the like constitute practical measures in this strategy, while adherence to guidelines which are understood and accepted by those concerned is essential.

It would be surprising if the level and balance of actions favoured at one time were always correct by reference to current attitudes and preferences as revealed by social and economic studies, much less that these attitudes were to remain unchanged. Such a fluid state of affairs cannot create confidence among those using the rules and it is in the interest of all concerned that owners should encourage continued enquiry into the topics concerned, the most critical of which must currently be considered to be the valuation of wildlife conservation outputs.

Control of afforestation is a further and usually more problematic issue. This is so because, in addition to the selection of the form of woodland to be created, questions arise as a result of the displacement of existing land uses. A new chapter in the control of land use appears to have opened with the enlightened form of land use planning formulated in Strathclyde Region (Goodstadt, 1992). Since forestry is not subject to statutory control by planning authorities, this strategic approach relies on voluntary collaboration of foresters, other land use interests and planners.

Apart from matters already covered by the various environmental guidelines of Timber Growers UK and the Forestry Commission, an important issue to be addressed is that of the contribution of private forests to public access. In aggregate, the potential of this use of private woodland is almost certainly of higher interest to the layman than many more recherché values. In the absence of the political will to legislate on the lines propounded by Shoard (1987) and of local authorities to press for access agreements under the 1949 National Parks and Access to the Countryside Act, financial incentives remain the principal means of achieving wider access.

Financial aspects

Taxation

Taxation is crucially important in encouraging owners to maintain forests and in discouraging or encouraging their extension. But this does not mean that exemption from tax (income tax is the relevant case in Britain and, among the other countries surveyed, Ireland) is necessarily to be favoured. One reason is that forests yielding revenues from wood cutting will attract a special value owing to the absence of income taxation of their net revenue. This obvious distortion will encourage tax evaders to acquire woodlands, though with a totally different motive from that applying in the cases of death bed purchases in the most extreme exploitations of the old Estate Duty regime or available after 2 years under the latest (1992) inheritance tax provisions in Britain. It is immaterial that such owners may have no special knowledge of, or interest in, forestry: the incentive for them to make the best return will certainly remain. But there will be greater fragmentation of wooded properties, with extra transactions costs, as so often rehearsed in the arguments surrounding the maintenance of larger holdings in relation to transfer taxes. The reason for the comparative absence of this type of transaction in other European countries which offer major tax concessions lies in the fact that purchases of land by outsiders are so restricted, a practice whose extension to Britain would be an outcome devoutly to be deplored.

The second and politically more important objection to the present arrangement is that any such provision is automatically regressive in a system which still purports to have a redistributive aim. Neither of these two criticisms is any different from the objections that could be raised, though rarely voiced publicly at the time (Price, 1971 is a stalwart exception), against a negligible or zero charge to income tax under Schedule B. Two successive wrongs do not make a right. Income tax is such an important, and accepted, charge on all other economic activities as well as on the forests of most European countries that the current British position on it is an issue that cannot be ignored in deciding forest policy.

There are sound practical reasons for taxation and provision for making the cost of approved operations tax-deductible. The position on many estates is that the provision of environmentally desirable improvements on farms can be dealt with by such deductions, but the same does not apply to the woodlands. The farmer using a machine to carry out a farming or conservation operation one morning cannot charge the same machine against the cost of establishing a plantation on his farmland the same afternoon.

The common feature of the typical forestry investment in Britain is that the individual gaining such income is rarely the person who invested in the establishment of the trees concerned. It is of course true that the original investor was, and continues to be, led by a generous government to expect that the stands he created could be bought or transferred at a value reflecting the fact that net revenue would not be taxed. There is, however, no logical reason for saying that because his father went to the trouble of creating the plantation in the expectation that his son would not be taxed, the son should be insulated from taxation on its income. There can be no contract between government and the investor safeguarding the post-tax cash flow from an investment producing income over many decades into the future.

Of course, the subject cannot be left there since the impact on confidence must be considered. There can be no doubt about the effect on planting of the announcement in March 1988 that forestry would no longer be subject to income tax. The reintroduction of forestry to income tax but without the operation of the favourable Schedule B rule (namely automatic assessment on that schedule with change of occupier) would accordingly be expected to have a marked impact. On the one hand, the change would put the investor in the same position as he stands in relation to any other investment opportunity, on the other, extra financial assistance must be provided if private afforestation is to be encouraged.

This matter of the appropriate income taxation of woodlands raises in acute form the dilemma of forest policy making already referred to elsewhere, namely that economic and social circumstances are liable to alter significantly within the life time of individual stands. The obvious method of resolving the current anomaly is to make forestry liable to income tax. This implies that expenses of approved operations would be allowable against other income and that income from timber harvesting would be taxable under Schedule D of income tax. No immediate change would be necessary in grant levels. The following advantages would flow from this change. Firstly, investment in new woodlands and in the maintenance and protection of existing woods would again become attractive to newcomers to forestry and existing estate owners. Experience teaches that the substantial scale of investment which is required in forestry is impracticable except through major capital sums provided by individuals and that the attraction of a reduction in tax liability is a considerable incentive. Although the use of allowable expenses appears regressive in such a long-term investment as a plantation, the desire to encourage private participation in afforestation unavoidably demands the use of the tax route. Secondly, equity requires that revenue-producing woodlands should not escape the tax net. Thirdly, grants may still be used to steer owners' actions. The effect of a given grant level will, however, be diminished by taxation at a marginal rate of 40%, and there may have to be some widening of grant differentials to provide effective signals.

The net effect on the Exchequer would be heavily dependent on the scale of tax-deductible activity generated and on timber prices. It is very likely that the adjustment would involve a small outgoing from the government if new planting were to increase from recent years' levels. Small flows in whatever direction do not mean that the change should be disregarded on *de minimis* grounds. Inheritance tax adjustments accepted in various Budgets have often implied quite small (downward) changes in government revenue.

Grants

As already noted, the majority of countries surveyed adopt the practice of paying grants based on a specific percentage of an agreed cost for the particular operation. It is likely that this basis for payment arises because the operations can be regarded as conventional ones and the applicant is viewed as a contractor carrying out a well-specified operation. There are two advantages with this method and one disadvantage. First, the degree of grant incentive can be more finely tuned to the region (as in Germany, Switzerland and Norway), to the individual's circumstances and to the site. Second, it is possible to reduce the extent of 'producer's surplus', that is, the profit that the low cost operator enjoys. This system is an incentive to making the best use of resources where there is a wide choice of investment opportunity, but there is no point in offering a grant rate that is higher than necessary to elicit the desired response. The government can become a discriminating monopsonist, and do itself good by being so. As it happens, there is much practical experience of the system in agriculture in Britain as pointed out by the Agriculture Select Committee (1989b). A preliminary agreement is needed on the cost of the operation at any given site before it can proceed, so that administrative costs are marginally raised on this account. The possible diasadvantage is that cost estimation may be slack. This may well be a concern in some countries but does not constitute a difficulty in ones with attitudes and practices broadly similiar to those found in Britain. In any case, use of the system accumulates cost data useful for checking.

Afforestation is not a significant activity within most of the countries reviewed here. This is probably the reason for the payment of similar rates of grant, whether set in absolute or percentage terms, for new planting and restocking. It is not at all obvious that restocking should be rewarded with a lower grant than new planting; the former is an obligatory action after felling, the latter voluntary. In addition, experience teaches that site for site the costs of restocking will often be higher than those of new planting.

Accounts show that the costs of administering forestry grant schemes in Britain, over and above consultation costs, amount to £27 for every £100 of grants disbursed (Forestry Commission, 1991b, p. 56). Site visits are increasingly concerned with detailed discussion of practices to be adopted and are necessary under any grant application system so that the requirement of a cost estimate does not add to the public expenditure on a field visit. There thus appears to be reason to investigate whether a change in the system to percentage rates, subject to maximum levels of absolute grant, could reduce total outgoings.

The basis of payment of management grants under the extended woodland grant scheme from 1992 was essentially compensatory, being designed to assist those whose expenditure on maintenance operations post-establishment was substantially altered by the removal of forestry from income tax. These grants could be reconsidered on the reintroduction of income taxation.

Extension

When the 1967 Forestry Act referred (S. 8 (b)) to the promotion of forestry, the activity in legislators' minds was not the education of the general public but primarily the training and information of the agents of forestry operations. This audience probably remains the most important one. The range of skills demanded of the growing number of investors and managers active in forestry places a higher premium than ever on competence. In these circumstances there is an enhanced need for extension services which are more readily available, correctly targeted and responsive to user needs. This task of promoting forestry can be carried out by a number of routes, including dependence on the market. The risk of such dependence is, however, that there will be such a deficiency of demand for the message that the outpourings of dedicated researchers will never be heard, let alone absorbed. It follows that expenditure devoted to research for a practical application has to be viewed along with expenditure on extension as serving one end. Desirably a balance should be determined whereby the returns at the margin are equal in the two activities. Economic assessments of this point are clearly difficult and a broad judgement may have to be made: a decision cannot however be shirked.

Two steps may be considered. The first is to establish within the forest authority a cadre of people whose sole concern would be with extension services to both the private sector and the forest enterprise. Their functions would thus be separated from those of research and enquiry in the various fields of biology, technology and economics. Instead they would be concerned solely with the transmission of integrated information to the field in both private and state forestry enterprises. They would be staffed on rotation from either the forest authority or the state forest enterprise, with a number also on secondment from the private sector. The second step would be to finance this forestry extension group by a reduction in other expenditure, most plausibly research.

The second purpose of extension, that of educating the general public, is emphasized in several countries, and the case for greater attention to this activity has already been noted in relation to the blandishments of single interest groups. The need is greater in Britain where the community is further from its roots in the land than elsewhere, and where forestry is an unknown among laymen. There is a risk that this activity may be seen as a public relations propaganda exercise. This attitude must be avoided. The facts should be displayed, and the arguments of detractors squarely faced. Several routes are open to use, often relying on local action. Talks to local bodies, participation by all woodland owners and not only state foresters, the creation of 'Friends' associations linked to particular woods or state forests are among the options.

Research

Total expenditure on forest research in Britain rose steadily in real terms over the decade to 1991. Since applied research, the predominant form of research in forestry, together with development work, produces the information transmitted by extension workers, it is appropriate to consider the desirable scale and organization of research.

Expenditure on research in the nine European countries surveyed may be compared with area, new planting, increment and cut for all ownerships combined. The indications are that in these measures, and ignoring the weights to be attached to unquantified outputs such as public recreation, hunting and conservation, Britain's research intensity is similar to, or slightly higher than, that of the smaller European countries as a whole, and all are very much higher than those of France, Germany, Sweden and Norway.

Whatever the total effort devoted to research, it is clearly important that the requirements of the private sector are met. Many countries have advisory boards to assess research information needs and to review programmes. Where there are substantial domestic wood-using industries, these also participate; they may in addition assist with finance. In Britain, while the scale of industry remains small relative to the area of forest, the quality and quantity of wood which will be harvested several decades from the present is determined by silvicultural decisions on planting and thinning today. It is appropriate for wood-using industry to be associated with growers in discussing the scale and balance of research conducted. Such a board would also include among its tasks advising the head of the forest authority on research requirements, reviewing the aims and achievements of research programmes of the principal contractors, fostering links between researchers and users, and developing new sources of finance. The establishment of a users' forum in 1992 by the Forestry Research Coordination Committee, and the creation in 1993 by the Forestry Commission of a research users' advisory group relating specifically to Forestry Commission research are to be welcomed.

Analysing and Monitoring Public Support and its Effects

Background

Forestry Commission expenditure in connection with private forestry in 1991 is estimated as shown in Table 18.1.

Allowing for expenditure by other government departments on research, grants, administration of grant schemes, etc. of at least £10 million, total public expenditure associated with private woodlands in Britain in 1990/91 was thus of the order of £40m. Such a large sum deserves a substantial effort to be devoted to its appraisal in terms of value for money. The following section notes a number of topics deserving attention.

Policy analysis

Although there is no strong tradition of policy analysis in Britain such as exists in the USA with the Brookings Institution, there is a concern with management efficiency. Policy analysis implies much more than this; it has two important functions. One is to improve the methods as well as the content of policy reviews, the second to improve the implementation of policy by assessing the merits of existing tools of policy and indicating

Table 18.1. Forestry Commission expenditure associated with private forestry.

	£m
Grants	16.458
Survey of private sector costs	0.120
Grant schemes: consultation costs	0.977
Grant schemes: other management costs	4.491
Felling licensing: consultation costs	0.137
Felling licensing: other management costs	0.433
Felling licensing: expenses arising from illegal felling	0.184
Sub-total	**22.800**
Additional items (estimated shares of totals)	
Share[a] (say 33%) of information and publications	0.669
Share (say 33%) of training	0.030
Share (say 50%) of plant health management	0.420
Share (say 50%) of other management costs	1.234
Share (say 50%) of research[b]	4.627
Sub-total	**6.980**
Grand total	**29.770**

[a] Author's estimates of share of total.
[b] Value differs from that shown in Chapter 5 owing to difference in accounting year.
Source: Forestry Commission financial statements, Forestry Commission (1991b).

desirable changes in policy tools. (Appendix III illustrates applications in the implementation field with special emphasis on extension.)

First and most important in relation to setting the agenda for policy reviews, the forest authority should be better informed of the wants of the country as a whole. Because of the sensitivity of economic analyses concerning policy, there is a marked dearth of published economic work on forestry with particular reference to objectives and goals. It is gratifying to note that this situation is changing through the work organized by the Forestry Commission (1991a). As with research on forest products, economics does not loom large in the in-house research programme of the forest authority. Work should therefore be commissioned in order to correct the imbalance, an aim recommended to the Forestry Research Coordination Committee in 1988 (Forestry Research Coordination Committee, 1989) but by no means fully implemented. Among the topics requiring attention at the level of broad policy objectives are:

1. Examination of the arguments underlying environmental ethics, and, within this subject, assessment of the implications of sustainability in all aspects of forestry.
2. Continuation of the research on valuation of non-market benefits, including studies of changes in government policy and in consumer tastes.

3. Analysis of trends in the markets and prices for wood products and roundwood.
4. Analysis of the costs and benefits of education of the general public in forestry matters.

Items **2** and **3** above cover a wide variety of topics with considerable calls for data. Thus, the issue of carbon storage demands, among other things, assessment of the influence of different species and silvicultural regimes on the value of woodlands and their production as temporary carbon stores and, in addition, prediction of the possible impact of a carbon tax on wood-processing industries and hence of market prices of products. Knowledge of the consumption of different forms of recreation and of factors influencing future consumption is poor. For several of the non-market benefits the lack of a mensuration of the output is as serious as the absence of unit values. In relation to measurable outputs, all categories of owner share responsibility for the provision of information.

In relation to programme analysis (narrow sense policy analysis) the forest authority's aim must be to assess and understand the reasons for the actions of owners in response to policy and particular implementation tools. Such understanding is a prerequisite to the improvement of existing policy tools or the introduction of new ones. The distinction between fields of enquiry serving broad policy needs as opposed to programme analysis is, however, arbitrary: candidates under the second heading which appear to deserve attention include:

1. Data on the characteristics of woodland owners, their length of tenure of woodlands, their resources, and objectives.
2. The rationales of new investors and existing owners in undertaking afforestation.
3. Owners' valuations of unsold output such as sporting and the management interaction with public recreation and wildlife conservation.
4. Timber cutting and marketing behaviour of different categories of owner.
5. Tax burdens, tax shifting and tax impacts on forest management.
6. Public expenditure and resource implications of using private or public agencies for particular functions, such as extension, training and research.
7. Effectiveness of the delivery of technical and economic information.

Research on each of these topics calls for data. Information on the amounts and composition of private woodlands expenditure and receipts is currently lacking. It is difficult to see how intelligent views of the success of private forestry policy can be formed without such information. Given annual public expenditure on all non-Forestry Commission woodlands of £40 m, of which two-thirds arise in connection with grant schemes, it is remarkable that economic surveys of the private sector are in abeyance. Simply monitoring the uptake of grant schemes aimed at establishment of more plantations provides a very limited view of owners' outlay and yields no information of use to policy development.

Of less importance but still relevant in this context is the deficient state of the British national income accounts for private forestry. The value of change in value of stocks in private forests in Britain is ignored although this is an essential ingredient of national accounts which currently seriously underestimate the value of economic activity in British forestry.

Since the value of information on this point of economic statistics is shared between private forestry organizations and the government, growers must be willing as in the previous sample surveys of income and expenditure to devote time to the completion of data[1]. The main beneficiary of more critical methods of assisting private forestry is the State and accordingly it should bear the brunt of the costs of financing both the surveys and the studies required.

The potentials of different forms of organization of the private forestry administration are relevant in relation to item **6** above. This work can only be carried out by modelling. Also, the performance of the chosen organization requires to be checked. The monitoring of administrative tasks within government departments and businesses is widely practised: standard costs can be estimated for budgeting and controlling administrative functions as they are estimated and used in controlling field operations. Where quantitative targets cannot be set, it is possible to set targets of duties to be carried out (Ministry of Forestry, 1991).

It is clear that the studies suggested are diverse and call for work in several disciplines. Much of the work outlined under broad policy analysis would be contracted out, whereas work on implementation studies might be carried out either by the forest authority or by contractors with access to official records. It is noteworthy that many of the studies recorded in Appendix III use cheap techniques such as telephone interviews: the cost of acquiring the requisite information by modern techniques need not be high.

Owners

Private woodland owners' collaboration with the whole panoply of forest authority functions is essential. The first need is strong support of the owners' organization. Without cohesion, owners will necessarily lose. A *laissez aller* attitude neither deserves nor wins respect from those in other

[1] The data on income should be more reliable and simpler to compile when British private forestry has matured to the point when it can organize the marketing of roundwood more successfully.

organizations influencing forestry, and private forestry in particular. In the face of an endemic indifference among some, it is difficult to suggest how greater enthusiasm can be engendered except through the organization demonstrating its value to both its members and the whole non-State sector.

A number of changes in the functioning of the owners' organization suggest themselves in the light of observation overseas. In the first place, study of the more successful of foreign owners' associations would be valuable. In particular, the relation between trading and representational activities needs to be understood owing to the implications for financing an association's activity.

Second, a standing body should be designated to keep policy aims and methods of implementation under review. This body would maintain close liaison with policy studies of the forest authority in order to ensure clear understanding of emerging problems, to initiate proposals and to provide relevant evidence on the implementation of controls, guidelines and incentive programmes.

Third, private growers should encourage and assist in the collection of data on the financial aspects of forestry. The demise of the surveys of costs and later of income and expenditure by teams who built up an enviable expertise and rapport with estates is regrettable. The owners' organization itself requires a firmer knowledge of activities and cost and receipts of the sector. The objectives of such data gathering would include: the monitoring of the financial position of forest owners, understanding of the treatment of resources, including management of forest capital, and the provision of an agreed data set for use in the consideration by both sides of the impact of changes in support methods and rates. The organization's members should therefore prepare or assist in the collection of the necessary physical and financial data. Part of this material would come, after a suitable time-lag, from any marketing function which the organization might undertake.

Fourth, if owners are to reduce their heavy dependence on government support they have to improve the industry's performance in revenue-producing activities. Bary-Lenger and Stordeur (1992), writing of the situation in Belgium, suggest that there are many actions that have to be undertaken if it is still to be worth owning private woodlands. In their case they consider the financial management of estates to be often archaic but remediable. In Britain, a more systematic approach to timber, especially sawlog, marketing is necessary, a field in which continental experience is strong.

Fifth, if owners are to be in a position to influence decisions stemming either from the British government or more particularly from the European Community, stronger links with other European growers' organizations are desirable. The single market represents but one step in the evolution of the European Community; agricultural policy changes have even more fundamental implications for forestry in the medium term. A joint input must be valuable in restraining the growth of unrealistic ideas on the wilder shores of Commission thinking. There would also be gains through learning how growers in other countries solve their problems.

Concluding Remarks and Summary of Main Recommendations

Woods may still in some quarters and at some times be regarded, as one Restoration peer put it, as excrescences of the earth provided by God for the payment of debts. But they do have other functions, as the woodlands of long-established estates especially demonstrate, which they perform very well. It is the newer forests, created at the nation's behest with the aid of public money, that have been the source of most of the present discontents. Policy cannot stay still; it has to develop to take account of the pressures to adjust the lay-out and silviculture of forests and woods to meet a multiplicity of purposes. In the process of resolving any changes in aims and goals to which private owners and their managers are to work in return for public money, it is important that all those engaged in developing policy deal honestly and fairly with the issues at stake. The present study provides some evidence from abroad, and ideas founded on those examples, which it is hoped may point to ways of achieving a sensible shaping of private forestry's future role.

Twelve principal recommendations arising from this study are as follow:

1. Policy review procedures should be developed to ensure both wider public participation and, as far as possible, cross-party agreement.
2. Increased and continuing effort should be made to inform all interested bodies and the general public on forestry matters.
3. Continued efforts to quantify and evaluate non-market benefits should be supported.
4. Consideration should be given to the reinstatement of income taxation of commercial woodlands.
5. Financial and management implications of relief from inheritance tax should be studied.
6. Consideration should be given to fixing grants as percentages of costs of the operation concerned.
7. Grant differentials be increased, once income tax is reintroduced.
8. Increased attention should be given to extension services to the industry.
9. Consideration should be given to the establishment of a forest industry research advisory board.
10. Studies in policy analysis should be carried out in-house and on contract.
11. A determined effort should be made by the owners' association to gain a wider constituency and influence.
12. Stronger inter-EC links, notably between private owners' organizations, should be forged.

REFERENCES

Agriculture Committee (1989) *Report: Land Use and Forestry*. House of Commons Session 1989–90. HMSO, London.

Agriculture Committee (1989) *Land Use and Forestry: Evidence of the Institute of Chartered Foresters*. House of Commons Session 1988–89. HMSO, London.

Bary-Lenger, A. and Stordeur, P. (1992) [Is it still worthwhile being a woodland owner today?] *Silva Belgica* 99 (3), 33–38.

Dewar, J. (1991) New planting methods, costs and returns. Paper 13, *Forestry Expansion: a Study of the Technical, Economic and Ecological Factors*. Forestry Commission, Edinburgh.

Forestry Research Co-ordination Committee (1989) Forest economics and planning research. *Information Note* no. 20. Forestry Commission, Alice Holt Lodge, Farnham.

Forestry Commission (1991a) *Forestry Expansion: a Study of Technical, Economic and Ecological Factors*. Forestry Commission, Edinburgh.

Forestry Commission (1991b) *71st Annual Report and Accounts*. HMSO, London.

Forestry Commission (1992) *The Forestry Commission of Great Britain*. Edinburgh.

Goodstadt, V. (1992) Constructing an indicative strategy: the Scottish experience. Paper to conference 'New Forests for the 21st Century'. Forestry Commission, Edinburgh.

Grove-White, R. (1991) Politics, databases and vegetation surveys. Paper to British Ecologcal Society conference 'The future of vegetation survey'. Typescript. Lancaster University.

HM Treasury (1972) *Forèstry in Great Britain: an Interdepartmental Cost - Benefit Study*. HMSO, London.

Hummel, FC and van Maaren, A. (1984) Policy formation. In: *Forest Policy: a Contribution to Resource Development*. Martinus Nijhoff, The Hague.

Lönnstedt, L. (1991) Economic consequences of the Swedish forest policy. Typescript.

Mason, D. (1989) Pining for profit. In: *Proceedings of Discussion Meeting on UK Forest Policy into the 1990's*. The Institute of Chartered Foresters, Edinburgh, pp. 118–121.

Ministry of Forestry, New Zealand (1991) *Annual Report*. Wellington, New Zealand.

Pearce, D. (1991) Assessing the returns to the economy and to society from investments in forestry. Paper 14, *Forestry Expansion: a Study of the Technical, Economic and Ecological Factors*. Forestry Commission, Edinburgh.

Price, C. (1971) The effect of tax concessions on social benefit from afforestation. *Forestry* 44(1), 87–94.

Scott, J. (1942) *Land Utilisation in Rural Areas*. HMSO, London.

Shoard, M. (1987) *This Land is Our Land*. Paladin Grafton Books, London.

APPENDICES

I

Population and Forest Statistics for Countries Surveyed

The statistics in Table AI.1 have been extracted from ECE/FAO (1992) 'The forest resources of the temperate zones: results of the UN-ECE/FAO 1990 forest resource assessment'. The values are presented as representing a consistent set of data although in certain text chapters the figures supplied by the forest authority have been used in preference.

Table AI.1. Population and forest statistics (All values are in millions of units).

Country	Population	Land area (ha)	Forest area (ha)	Closed forest (ha)	Growing stock	Net increment	Fellings
					(volumes in m^3 overbark[a])		
UK[b]	57.41	24.09	2.38	2.21	203[c]	10.1[c]	8.1
Belgium	9.84	3.03	0.62	0.62	90	4.46	3.33
Bulgaria	9.01	1.02	3.68	3.39			4.76
Czechoslovakia	15.66	12.54	4.49	4.49	991	31.0	20.2
Denmark	5.14	4.25	0.47	0.47	54	3.52	2.29
France	56.44	54.33	14.16	13.11	1,742	65.9	48.0
Germany	79.88	34.93	10.74	10.49	2,674		42.6
Hungary	10.55	9.21	1.68	1.68	229	8.23	6.06
Ireland	3.50	6.89	0.43	0.40	30	1.57	1.41
The Netherlands	14.94	3.39	0.33	0.33	52	2.39	1.30
Norway	4.24	30.69	9.57	8.70	571	17.6	11.8
Poland	38.18	30.44	8.67	8.67	1,380	30.5	27.3
Sweden	8.56	40.82	28.02	24.44	2,471	91.0	57.5
Switzerland	6.71	3.98	1.19	1.13	360	5.82	5.3
Europe	564.9	550.4	95.0	149.3	18,509	576.7	408.3
USSR	350.5	2,217.7	957.0	770.2	51,925	733.6	543.8
USA	250.0	913.7	296.0	209.6	23,092	619.6	497.9

[a] Figures relate to exploitable forest, hedgerows excluded.
[b] Values for GB forests approximately 3% lower.
[c] Growing stock and increment values may be underestimated.

THE EUROPEAN SINGLE MARKET

INTRODUCTION

As noted in Chapter 4, the single European Act (SEA) marked the most significant change in the aims and procedures of the Community since its formation in 1957.

This Appendix sets out some of the main elements of the economic arguments for changes if the Community is to live up to the name 'common market', and draws attention to a number of initiatives that bear, or could bear, on forestry over time.

THE CECCHINI REPORT

The Cecchini Report (1988) represents the most thorough attempt to quantify the outstanding gains to be reaped as a result of the removal of non-trade barriers[1]. The title 'Common Market', Cecchini points out, is a term used with growing embarrassment and decreasing accuracy. It was recognition of this that led the European summit of 1985 to endorse Commission proposals to 'complete the internal market'. Three broad types of barrier are distinguished: physical (such as customs controls), technical (such as different standards and business laws) and fiscal (different VAT rates and excise duties). An informative analysis of the results of a questionnaire completed by 11,000 businessmen is presented in Table AII.1. This shows the ranking among UK and all EC12 businessmen of the importance they attached to eight factors.

[1] As J. Delors in his foreword to the report puts it 'a large market . . . gives a unique opportunity to our industry to improve its competitivity'. Mayes (1990) draws attention to the word 'our' in this quotation and to studies which have shown how gains to the Community may not be achieved without losses to the economies of other regions.

Table AII.1. Ranking of importance attached to various barriers.

	UK	EC12
National standards	1	2
Administrative barriers	2	1
Physical frontier delays and costs	3	3
Government procurement	4	8
Freight transport regulation	5	6/7
Commericial law	6	4
Capital market restrictions	7	5
VAT differences	8	6/7

The table shows a substantial degree of agreement between the views of the whole set of returns and those of British businessmen.

Cecchini's analysis of the potential gains from removal of all internal barriers is comprehensive; it yields very impressive numbers. As far as is known, no detailed assessment of gains in the wood and wood products sector has been made. Calculations by sector are of course liable to be partial, since some of the gains in fields such as liberalization of financial services are of their nature ones that can be evaluated sensibly only across the whole of commerce. The microeconomic part of the calculation sums the increase in consumer's surplus, that is the value above the market price that consumers would be willing in principle to pay and which increases as prices fall, plus the net change in producer's surplus that producers would have as a result of decreases in their input costs but diminished profits from more active competition. The gains considered were to be obtained from:

1. Economies of scale, derived mostly from restructuring.
2. Reductions in price levels and profit margins, calculated via observable price variation for particular goods and assuming a reduction in variation towards the lower end of the range, these two elements alone totalling 4.3–6.4% of the EC's GDP in 1988.
3. Market dynamics leading towards the more rapid adoption of technologically advanced processes.
4. Success of business strategies for the European home market through easier entry, standardization not only of products but also of processes.
5. Improved financial reporting and information flows generally.
6. Removal of border controls.
7. Opening up of procurement markets.
8. Liberalization of financial markets, recognized to contain many cartelistic features.
9. Other supply side effects.

These gains totalled some 4.5% of GDP attainable over the 'median' period, though this could, it was believed, be further enhanced by improved demand management, leading for example, to higher investment.

The report concluded that the guidelines to success involve responsiveness on the part of business, active competition policy, some wealth redistribution, and success in the application of general economic policy. The relevance of the Cecchini analysis for forestry is obvious in relation to the first two of these factors. The third factor of wealth distribution effects may well be important, since more rapid growth in some industries leaves certain regions at a disadvantage and forestry may be called upon to assist the social adjustments that are required.

A whole library of studies[1] has shown the magnitude of the so-called static gains from the formation of a customs union. These arise from trade creation, trade diversion and trade suppression. They are termed static because they refer solely to the change from before the adoption of a single external tariff to the position after the change. They all assume perfect competition in the markets for goods. Figures for the EC6 in the mid-1970s suggest that the gain from introduction of the common external barrier was of the order of 3% (60 billion 1986 ECU relative to a total GDP of 2000bn 1986 ECU). But the important point to note is that they are likely to be substantially smaller than the dynamic gains. These arise mainly on the supply side of the economy. The main elements are: (i) the stimulus to greater efficiency of existing productive capacity; and (ii) the stimulus to the rate of investment that is engendered.

The Commission has achieved rather uneven progress in the application of the laws which are aimed at the promotion of competition, with more success for instance in relation to manufacturing than road haulage or shipping. Some of the treaty's Articles are reviewed because of their importance for trade and industry.

Article 85 of the treaty is concerned with agreements between firms, or concerted practices, which have as their result the prevention, distortion or restriction of competition. An interesting application of this Article quoted by Nevin (1990) was the successful prosecution by the Commission of an international trade association in what is known as the Paper-Machine Wire Case of 1978. Members of the association provided a central secretariat with details of their prices and other trading conditions, and in return obtained similar information from their competitors. It was claimed that the purpose of these exchanges was to prevent members being given false information by prospective customers about the terms they had been offered by other suppliers. Unsurprisingly the Commission was not convinced, especially as it was conceded by the association that this mutual exchange of information was not extended to potential buyers. The arrangement was dismantled as a result of the Commission's action. Other arrangements covered by Article 85 include: cartels, market-sharing agreements, the application of different

[1] A highly competent and readable account of the economic processes involved in economic integration is given by Nevin (1990).

conditions of sale to different purchasers of the same product, and conditional sales in which the purchaser is required to buy unconnected products when making purchases.

Article 86 concentrates on collusive arrangements between independent firms and pays particular attention to cases of firms which have 'a dominant position' in the market. These cases are difficult to define and pursue.

Article 92 provides for action to be taken where state aid given to enterprises is deemed incompatible with the Treaty of Rome. Nevin suggests that competition has been far more restricted by government intervention by way of subsidies or less obvious means than by any private monopolies or cartels. Under the Article a point of principle was gained in the Cassis de Dijon case of 1978. In this, a German merchant wishing to import a French liqueur, Cassis, was prevented by the German authorities on the ground that under German law the Cassis did not contain enough alcohol to be classified as a liqueur. The European court ruled that a government could not prohibit a commodity the sale of which was allowed in another Member State unless its import was likely to be injurious to human health or consumer protection. While the Commission has attempted to apply the general rule to trade in the Community, governments have proved inventive in finding new ways of restricting imports from other Member States or indeed elsewhere.

Plant Health Control

Since 1980, harmonized plant health controls have applied to imports to and trade between EC Member States. The abolition of internal frontier controls with the advent of the Single Market implies the abolition of plant health checks when material is moved from one EC country to another. In recognition of the importance of retaining control over the movement of certain forms of wood and forestry materials for purposes of protection from pests and diseases, plant health checks are retained but are focused on the place of production (Forestry Commission, 1992). A plant passport will be required for movements of designated materials: the main point of the document being to show that material in the particular consignment is free from all relevant pests and diseases. The administrative arrangements impose an inevitable cost on the forestry sector: it will be important to ensure that they do not become too cumbersome.

Standards[1]

So far as standards are concerned, the principal rationale for harmonization lies in the fact that firms want larger markets but national regulations on products and their standards impose a contrary logic. Thus if firms wish to

[1] I am indebted to Mr John Sunley for factual information incorporated in this section.

attack new markets they must adopt suboptimal sizes of plant. The survey of businessmen noted in Cecchini, after appropriate weighting, showed that the order of importance of technical barriers led with cars, products of electrical engineering, products of mechanical engineering, with leather goods far lower. It must be assumed that timber products are well down the list.

The European Standards Institute was formed in 1961, consisting of CENELEC (the European Committee for Electrotechnical Standardization), dealing with electrical goods, and CEN (Comité Européen Normalisation) dealing with all others. Following slow progress during the 1960s and 1970s and as the move towards the need for a unified market gathered strength, the Council of Ministers resolved on a new approach in order to overcome the difficulty of standards for new goods becoming slowly established. The new approach of 1985 called for:

1. Mutual recognition of products (the notorious Cassis de Dijon case clarified the position on this).
2. Strengthening mutual information procedures (the Commission had since 1983 to be informed of new regulations and standards).
3. Selective harmonization implying on the one hand a limited number of directives to cover essential safety requirements and, on the other, the aim of establishing European, that is, wider than EC, standards to which industry subscribed using CEN to draft technical specifications.

CEN thus includes EFTA members, a particularly important and positive virtue in the case of wood and wood products. The aim set out was that EC standards established by CEN should adopt as far as possible other international standards organizations' (ISO and IEC) standards. The work of constituent committees has been carrried out by national standards institutes (the British Standards Institute in the UK case) who work with EC technical committees.

The harmonization of standards for timber and timber products centres on materials used in construction. The Construction Products Directive aims that all constructions should meet requirements in the following areas: mechanical resistance and stability, fire safety, hygiene, health and environment, safety in use, protection against noise, and energy and heat economy. Work on timber and (mechanical) wood product standards was divided among the following technical committees:

1. TC38 on durability of wood and related materials.
2. TC103 on adhesives for wood and derived timber products.
3. TC112 on wood based panels.
4. TC124 on timber structures.
5. TC175 on round and sawn timber.

In addition, TC250 SC5 has the major task of drafting a Eurocode on timber design. Each technical committee has used a considerable number of working groups to progress the work: for example that on panels has six. Provisional standards agreed by a technical committee have to go for final approval to the main CEN. Of the 'commodity' TCs, it is likely that progress will be slowest for TC 175 with target dates for many standards running into 1994. Two of its working groups are of particular concern to foresters: WG2 on sawn timber and WG4 on round timber. Progress on sawn softwoods has been difficult, reflecting historic differences between the conventions on preferred sizes among suppliers from north and central Europe respectively.

One cause of slow progress generally is the multiplication of standards, with a total of 201 (as of October 1992) to be handled by the five timber TCs. This large number has a number of consequential effects. For example, some cannot proceed without measurement and definitional standards being agreed; it would be absurd were any of the different working groups to develop their own. In addition problems will appear as standards are issued. Because new standards will cover a subject in a different way from the existing British standard (BS), withdrawal of the BS would lead to breakdown of the system until all new standards have been promulgated, a process which in some areas is expected to take two years. It is difficult to avoid the conclusion that the final goal of encouraging freer trade is far from the minds of those more interested in regulation for its own sake.

Progress in TC 175's working group on round timber will be of some importance to foresters because of trade in logs. Also an important test will be the implementation of new standards for sawnwood in relation to North American timber, since shippers there consider that CEN creates a barrier to much of their material.

Other Harmonization Steps

Taxation

In the 20th century, domestic taxation has come to be considered an important tool of macroeconomic policy. The issue of the range of control exerted by government over the level of economic activity has been an important element in the debate over moves towards a common currency. The first loss of economic sovereignty, the acceptance of the common external tariff, was of course an essential part of acceeding to the Treaty of Rome. A second was the shift in the United Kingdom from purchase tax and in Germany from a broadly standard turnover tax to value added tax. The arguments for tax harmonization are often subtle and the need for harmonization is by no means always obvious. First, the case against turnover tax may be used as an illustration. Because the tax is only borne on sales by one firm to another or to the final consumer there is a strong incentive for firms to merge, even though this may not be sensible on grounds of economic efficiency and may lead to the growth of monopolies. Second, firms obtain a tax advantage from producing their own components instead of importing them. Here again a distortion arises from what would otherwise be the financially most attractive course were there to be no turnover tax. Value added tax has the merit that it does not produce such distortions. However, other considerations arise in relation to the harmonization of VAT in the Community.

Value added tax

It might seem obvious that the free movement of goods, one of the key objectives of the Treaty of Rome, must imply the harmonization of VAT rates. This proposal to remove an internal fiscal barrier has met with strenuous opposition not because it is regarded as wrong in principle but because countries have strong political attachments to their own rate structures. Thus in Britain the government has long adopted the practice of zero rating food and children's clothing. On the other hand in Greece and Italy, where the collection of income tax is fraught with difficulty, heavier dependence on indirect taxes as a source of revenue is illustrated by their use of high rates of VAT, respectively 36% and 38%. In effect a pragmatic attitude towards VAT harmonization has had to be adopted by the European Commission. An interesting twist in the argumentation deployed by Britain in 1988, reflecting the fashion of the time, was the idea that market forces could be relied upon to bring about whatever degree of harmonization of rates was needed to overcome problems of distortion and cross-border shopping (Knust and Jenkins, 1989). This is hard to believe at the low end of the range but there is no reason why maximum rates should not be left to national decision since here the forces of international competition could be relied upon to lead to convergence. The target is now spoken of in terms of 'approximation' and it remains uncertain at what rate there will be this move to convergence either by regulation or through market pressures leading countries to change their rates.

It is assumed that there will indeed be convergence towards the rates proposed by the Commission, namely a standard rate band of 14–20%, and the differential values implied are not judged significant in the assessment of the relative attractiveness of investment in different locations.

Direct taxes

INCOME TAX

The Neumark committee set up by the Council of Ministers, whose report appeared in 1962, wrote at a time when labour was relatively immobile and it was therefore judged that no great benefit would be likely to flow from the harmonization of personal tax regimes. This conclusion might be judged differently today. Indeed it is this response to the market that may again prove decisive in moving member States towards convergence. There is a further set of considerations, however. Governments, as already noted, depend to differing degrees on direct and indirect taxes for their revenue. They also use fiscal methods to a greater or lesser extent to control demand and accordingly may be presumed to be unwilling voluntarily to relinquish control of their income tax rate structure. The advent of the single currency would have major implications for taxation as it would for monetary controls (that is, interest rates). For, as Lomax (1989) points out, governments could take resources from the rest of the Community by running deficits. On this argument, centralized control of national tax policies becomes increasingly necessary.

Meanwhile a question remains about the effect on competition of differential rates of income tax. Neumark considered that variations in rates of income tax would have little effect on the allocation of resources because labour would not migrate in response to tax differences. New evidence of the relative immobility of labour has appeared from a study (London Economics (MES) Ltd, 1992) which suggests that large wage and salary differentials will persist for at least 20 years despite the abolition of barriers to the movement of capital and labour. It is argued that the main reasons for convergence over the past decades of wage rates among member States has been changing employment patterns, for example the marked shift out of agriculture in several countries, and not migration of labour.

Corporation Tax

A 1975 draft directive recommended that corporation tax should be fixed within a band of 45–55%. In 1989 corporation tax rates ranged (Knust and Jenkins, 1989) between 34% (Luxembourg), 35% (Spain and Britain, with its lower rate for income below £150,000, and The Netherlands with its higher rate of 40% for the first gld. 250,000) rising to 50% in Denmark and 56% in Germany (although this was reduced to 50% in 1990). Although differentials may be expected to have an influence on the location of businesses such as those investing in wood processing in the 12 with the accession of Nordic countries to the Community, wood-using industries would hardly be expected to be affected greatly since they are hardly among the more mobile. In any event governments will increasingly take account of the same argument concerning the virtues of convergence as noted for personal income taxation.

Some progress has been made in agreeing consistency of taxation of dividends; this is a desirable step since the ways in which investible funds move are heavily affected by differential treatment of dividends paid in different countries. Also, capital gains tax laws differ among countries, so that, for example, a Dutch company would not be taxed on a gain on disposal of an asset held in another member State whereas a British o· German firm would be so liable.

Establishment of professsions

It is clear, as the British Department of Trade and Industry has put it, that the freedom of movement for people to work throughout the Community is among the essential rights established by the Treaty of Rome (quoted by Arkell, 1989). But, all member States restrict the performance of some professional activities to holders of qualifications granted by national bodies. Certain directives have ensured that training will be harmonized for specific professional activities and mutual recognition of higher educational diplomas (meaning degees and diplomas) accepted. So far the only professions with these arrangements for members to practise in any member State without being required to undergo any further examinations are architects and those engaged in health care (doctors, dentists, nurses, midwives, vets, pharmacists) under Directive 86/457.

A 1988 directive (89/48) on a general system for the recognition of higher education diplomas which came into force on 4 January 1991 aims to make it easier for other professionals to practise anywhere in the Community. The directive applies both to professions self-regulated by chartered bodies and to those regulated by government. There are no plans by the Commission to set up single Community qualifications. The directive has stimulated awareness among professionals, such as surveyors, of the possibility of increased competition in their home markets and of the potentialities of new markets elsewhere in the Community. It is unclear how rapidly the spread of professional firms into other countries will occur. In the case of forestry, where local knowledge and experience of the physical, social and traditional aspects that are encountered in practice loom large, it is likely that progress will be slow. The small volume of business for professional forestry advice must also be a strong deterrent to the opening up of this market. But the position may be different for tree planting and care in construction projects.

Financial services

Although major changes are impending in the valuable financial services market these are more likely to influence business in other member States than the United Kingdom. A major series of restrictions to borrowing exists; if dismantled the gain promises to be very large indeed. Walters (1990) has drawn attention to the cartel structures which are characteristic of banking in France and Italy. Since 1990, German insurance companies, which probably control more than 70% of long-term savings, were not permitted to buy non-Deutschmark denominated assets. In addition they could hold only 5% of their portfolio in equities, necessarily mark equities. In Britain too there will be rewards. For example, banks will have to look to their competitors abroad; some changes made in 1991 can be considered precursors of this more responsive mode to the needs of the customer.

Road transport

Of other activities with a possible bearing on forestry, road transport is important. Progress in opening up this industry to wider competition has been extraordinarily slow. Little liberalization of transport rules has been achieved and no progress made on agreeing how cabotage (the transport of goods within a country by foreign hauliers) should be handled. A limited amount of harmonization (e.g. restriction of hours of work, introduction of tachographs to monitor this aspect, road worthiness tests and driver training) has been agreed. The aim of a 1988 Regulation (88/1141) is to abolish all quantitative restrictions on international goods transport by road between Member States. There are substantial rewards to the breaking down of barriers in this field: for example the avoidance of dead running and the benefits of increased competition from other countries' hauliers. It is expected, however, that cabotage will remain, indeed it seems likely that it will be legalized (Dawson, 1989).

References

Arkell, J. (1989) The professions. Chapter 7. In: Clarke, W.M. (ed.), *Planning for Europe: 1992*. Waterlow Publishers, London.

Cecchini, P. (1988) *The European Challenge 1992: the Benefits of a Single Market*. Wildwood House.

Dawson, R. (1989) Transport: road haulage and rail. In: Clarke, W.M. (ed.), *Planning for Europe: 1992*. Waterlow Publishers, London.

Knust, J. and Jenkins P. (1989) Fiscal harmonisation. In: Clark, W.M. (ed.), *Planning for Europe: 1992*. Waterlow Publishers, London.

Lomax, D. (1989) Towards a common currency. In: Clarke, W.M. (ed.), *Planning for Europe: 1992*. Waterlow Publishers, London.

London Economics (MES) Ltd (1992) *Will the Single European Market Cause European Wage Levels to Converge*? 91, New Cavendish St, London.

Mayes, D.G. (1990) The external impact of closer European integration. *National Institute Economic Review* November, 73–85.

Nevin, E. (1990) *The Economics of Europe*. Macmillan, London.

Walters, A.A. (1990) *Sterling in Danger: the Economic Consequences of Pegged Exchange Rates*. Fontana/Collins, London.

APPLICATIONS OF POLICY ANALYSIS

METHODS OF ANALYSIS

The main purpose of policy analysis is the assessment of the effectiveness of policy measures. This appendix is not intended as a critical appraisal of this field of work, but as an introduction to a valuable tool of analysis. Illustrations are given of the kinds of elements and possible analytical, largely economic, techniques that are likely to be relevant in the study of owners' responses to policy initiatives.

Studies assessing the impact of government policies can be conducted at a number of levels of sophistication and generality. At one extreme is the use of forest sector models, that is ones which link forestry and wood-processing industry in a quantitative manner. Solberg (1986) discusses this application. However, these models, such as the major international exercise produced by the International Institute for Applied Systems Analysis (IIASA) (Kallio *et al.*, 1987), are large and complex, requiring huge data sets, and are not yet accepted by policy makers as being useful. This is partly because the number of policy makers 'thinking big' is small. Such global models do, however, stand, along with the econometric models referred to below, behind the projections of future prices of wood products and assortments which allow assessments such as that by Arnold (1991) to be made. Solberg (1986) noted that there is one feature which is a necessary element of their composition. The model must of its nature cover all the physical and economic interconnections, at least those that are included in transactions. This means that there is no risk that secondary or 'indirect' effects of a particular policy step will be overlooked, since all relevant variables and relations are included in the model. There remain two deficiencies of present models in this field. The first arises from the fact that they are closely linked with wood processing industry. Its interests are normally concerned with the short and medium term, say up to 15 years ahead, with the emphasis on the forests' prospective supply of wood over that period. Issues such as afforestation and reforestation which can typically only influence wood supply some decades ahead are not of great relevance, except in a broad political sense, to decision makers in the wood processing industry.

The second deficiency, and one shared by the econometric approach, is that consumption and production of goods or services such as informal recreation, which are not market transactions, or more generally are not mediated in money, cannot be adequately handled by such models. Some arbitrariness is necessarily introduced.

The alternative to models such as the IIASA *oeuvre* is the modelling of intra-industry supply and demand relations. This work has been progressed most assiduously and competently in the United States; Adams and Haynes (1980) have described the model developed for the US Forest Service's periodic timber trends studies which form part of the agency's decennial reviews of renewable resources, the latest use of which is represented by Haynes (1990) giving projections of wood and wood products supply, demand and price to 2040. Since the requirement in such studies is to look a long way ahead at the prospective supplies of wood products to the US economy and the prices of wood products and hence those of trees, this work requires projections of wood supply many decades ahead. In turn this requires a relationship to be assumed between factors influencing new investment in forests, such as stumpage price, and the resulting stock of stands. The relations devised to connect the response of private woodland owners to those prospects are necessarily broad brush. They can of course be made to include the effects of restrictions on operations such as harvesting on certain terrain.

The purposes served by such studies are much wider than the question of the effectiveness of government assistance to owners but constitute an essential background to much of it since if a given policy instrument affects supply and hence price, this factor has to be taken into account in any evaluation. On a small scale such as the Ringstad scheme for planting in the west of The Netherlands described in Chapter 9, it is highly unlikely that stumpage prices would be affected, but if a large programme of afforestation were in prospect, it would be irresponsible not to establish the likely impact on price of the major shifts in supply that must follow.

In order to investigate the ways in which owners respond to incentives, it is necessary to work in more detail. Desirably the assessment should be undertaken in relation to particular groups of actors since one group may well, and almost certainly will, behave differently in response to a given scheme from people in a different region, income group or occupation.

The principle of estimating the relevant parameters can be shortly stated: it is to establish a quantitative relationship, that is a function, which assesses change in the target variable, such as hectares planted, in terms of policy variables such as the level of planting grant, and on the other hand the owner's goals, the character of the forest holding and the general economic environment. It is a feature of the modelling work referred to and most of the studies of programme evaluation described below that they use regression techniques of one kind or another. The reader should be warned of the pitfalls of uncritical acceptance of the results presented by writers.

The studies described here concentrate on two subjects, first the impact of government extension services on forestry or, in one case, forest products industry, and secondly owners' decisions on cut, whether influenced by government action or not. The review draws heavily on five main works: Royer and Risbrudt (1983), de Steiguer (1983), the set of papers edited by Järveläinen (1981), the set edited by Tikkanen (1986) and a group communicated at the meeting of Division 4 of IUFRO at its XIXth Congress in 1990.

Extension, with Special Reference to the United States

The survey relies mainly on American material because the bulk of the policy analysis in this field has been carried out by American workers. The scale of US forestry is immense and its variety huge. To judge by the media attention they gain, from spotted owls to bans on certain silvicultural practices, it might be thought that federal forests were the most important types in the country. But federal and other public forests are actually in the minority by area and wood increment (Table AIII.1).

Non-industrial private forests (NIPF) not only occupy 57% of the total forest area but they account for 55% of the growth and, because the cut is relatively low, for 57% of the current addition to growing stock.

Their management is influenced by government intervention of one kind or another, but taken by and large on nothing like the intensity found in Europe. Thus, in some States there is almost no regulation of forest management practices, whereas in others, such as California, regulations embrace all operations. Substantial tax reliefs are allowed and these vary considerably between States. Both Federal and State forestry departments provide

Table AIII.1. Ownership of US forest resources.

Category	Area (m ha)	Growing stock (m m³)	Increment (m³)	Cut (m m³)
Non-industrial private	112	9,880	350	232
Private forest industry	29			
Federal	34	11,530	288	234
Other public	21			
Total	195	21,410	638	466

Source: Haynes (1990).

assistance programmes. The greatest contribution comes from the States with the bulk of this expenditure on fire protection, and support of silvicultural measures the next largest item. Cubbage, *et al.* (1993) provide a magisterial overview of public assistance to private owners as part of their comprehensive study of forest resource policy in the United States. A feature which stands out from their description of financial and technical assistance is the plurality of sources of public input via different departments' programmes (*anglice* schemes).

The pervasive concern of most commentators on national forest policy towards private forests is that their management should be improved. The arguments are never wholly clear, as indicated by Clawson (1979) in his monograph devoted to NIPF ownerships in the United States, in which he leaves until the last page, and then unanswered, the fundamental question 'Does the national security or national well-being require a volume of timber or other forest outputs that will not be forthcoming from the operations of the private market?' The usual arguments in favour of public assistance have long been paraded in the United States. Attention has been concentrated on the gains in national income from increased supply, some of which comes from programmes such as for fire and insect control which also yield external economies. Curiously little attention has been paid until recently in the arguments about the necessity for assistance to NIPF owners to the concept of market failure in relation to investment for long-term benefits that are uncertain. Instead, the emphasis has been on wood production, and the cost of this relative to current price, to costs in publicly owned forest and to the prospective price of stumpage. The last line is strongly developed in the latest timber trends study (Haynes, 1990). This incorporates a joint determination of price and supply and in relation to supply has the benefit of some detailed guidance on the investment opportunities for different ownerships, regions and operations. As noted in Chapter 16, the efficiency argument for public intervention to increase wood supply, and thereby reduce price, has been explored by Boyd *et al.* (1988) who conclude that there is no case. Other important studies have concentrated on forest owners' behaviour by attempting to identify the proximal factors influencing their actions with the aim of predicting how they may respond to changing economic or other circumstances. Binkley (1981) in a paper on New Hampshire non-industrial private owners shows how the probability that they would cut their forest was highly dependent on standing tree prices, with farmer owners being twice as responsive to prices as non-farmers.

American practice in the use of the word 'extension' is highly specific, usually implying education by varied more or less informal methods (Krygier and Deneke, 1983). Technical assistance is the term reserved for on the ground advice, normally one-to-one with the forest owner. As observed in the discussion on Sweden, there is a range of target audiences and hence of activities supplying technical information at various levels to them. Here extension is considered *sensu lato* and the audience will be taken to be those engaged in the forest industry.

Although the number of forestry consultants practising in the United States doubled between 1976 and 1990 to a total of 2500 this is not a large number when compared with, for example, Britain's Institute of Chartered Foresters figure of 110 (although this figure includes a proportion emphasizing arboricultural subjects). No complete census of public servants providing an extension is available, but some idea of the scale of the activity can be gathered from the figure for the North Region of the US. In 1978, 500 service foresters, or 275 full-time equivalents, were employed (Gansner and Herrick, 1980): this number should be seen against the total private forest area of 55m ha, including the relatively small share (13%) of forest industry timberland, and a low intensity of working overall.

Service foresters and officials of similar status are employed by the Soil Conservation Service, the Cooperative Extension Service and the associated colleges and universities responsible for land management, State Forestry Commissions, and the Federal Private and Cooperative Service. Total expenditure was $17.6m in 1976 (USDA Forest Service, 1982). In the same year Federal and State spending on forestry research was $121m. Since the early 1980s, as a consequence of the Reagan Presidency, extension programmes have decreased, though it is difficult to say by how much. There was an expansion of evaluation exercises on activities in support of private and, to an extent, all forestry from the late 1970s through to the mid-1980s. Partly encouraged by the IUFRO initiative noted earlier on the subject of policy analysis, this was largely stimulated by the anxiety of managers of leaders of Federal programmes to maintain their momentum against the prevailing coolness of the Administration towards government expenditure (Cubbage *et al.*, 1993). This was also noticeable in the research field. Federal expenditure on research fell by 25% in real terms to 1986, but non-federal expenditure increased by 30%, compensating in part for the Federal decline (Haynes, 1990). The concern led, along with other pressures, to major efforts on research evaluation culminating in meetings in 1984 and most recently 1989. Since the 1986 meeting on policy analysis, the effort in this field has been sparser but the quality of the debate sharpened by moving on to the rationale for government support of private forestry.

The majority of studies have used cross-sectional analysis of substantial numbers, usually hundreds, of forest owners interviewed to provide evidence on the relationship between the operation carried out, the availability or use of promotion, extension, or other guidance information and other variables considered relevant. Three techniques have been used in attempting to assess the value of extension. These are commented on in turn.

Cross-section analysis

Cross-section analysis has been widely applied to data for a sample cross-section of owners or operators with a variable included to represent extension input, usually assessed by presence or absence. Examples of the approach include the following.

1. Evaluating an extension programme for sawmillers designed to improve sawnwood recovery, Risbrudt and Kaiser (1980) regressed recovery on several variables including the interval between first and second visits to mills. The improvement in recovery appeared to be significantly influenced by the programme: the gain being 4.185% in the year following a study. The value of the increased recovery, adjusted for downward pressure on sawnwood prices, showed a high return on the programme's cost.
2. Young and Reichenbach (1987) developed the classification system noted in Chapter 3 on categorization of owners which distinguished those intending to carry out cutting from those not so intending. The relevance of their system for promotion of cutting was that if the aim of policy was to promote restocking after clear felling by means of a grant scheme, such schemes would be pointless for the 'non-intenders' since they did not wish to supplement their income by cutting wood. This led the authors to suggest that the need was to change the attitudes of such owners. This would in turn have implications for the message that extension foresters should convey, for example to increase the association in the owner's mind between cutting and acceptable practices such as providing firewood, or diversifying the habitat for wildlife.
3. A function of extension workers in the United States, which in many other countries would be looked at askance as being the role of others, is that of price reporting. Marty (1983) mentioned that the service forester acted 'as a trusted intermediary in timber sales and practice contracting'. Timber Market South is a publicly supported price-reporting service. Hyde and Newman (1990) found that price variation had been reduced in the years following the introduction of the service. The gains from greater stability were estimated and found easily to exceed the costs of the service. Gray's (1990) study of owners in New York State, although less rigorous in terms of analytical technique, indicated that their search activities for markets were limited and the quality of sources consulted poor.
4. Boyd (1983) used data for a random sample of owners as the basis for regressing the probability of harvesting on a number of factors including income, education level, area of forest, whether farmer or not, together with a technology factor simulated by whether or not the owner had received technical advice from a forester. So far as harvesting was con-

cerned, the results indicated that the last variable had a strong positive impact on the decision to cut, and roughly 50% stronger than knowledge of the availability of a grant towards restocking. The same paper pursued the matter of owners' interest in restocking and found that on the smaller number of holdings where cutting took place, the impact of technical advice was similar to that found with the cutting decision.

5. Royer (1987) investigated the determinants of restocking behaviour among 251 Southern forest owners. He was able to show that knowledge of one incentive, namely grant aid, was less influential than knowledge of tax credit and amortization rules.

Two Finnish studies which included extension as a possible explanatory variable may also be noted.

1. Järveläinen (1986) used data from the continuing data source organized in Finland to supply information on a large number of ownerships. Interviews were held in order to collect information on the use of extension services and regressions of the cut were then run on ownership and holding characteristics, timber sales motivation and use of extension services. Though there was a positive association between extension and the use of the forest's allowable cut, this was judged not to be especially strong. However, the result cannot be diminished by this modest assessment and, in particular, on account of the incorporation of motivational factors which are liable to be influenced by the use of extension services.
2. Karppinen and Hänninen (1990) in a study reviewing the use of the available allowable cut similarly drew attention to the uncertainties surrounding the extent of or indeed existence of causal relations between policy variables such as extension or grants and the target variable such as planting. Those owners who used extension services could be selected according to many factors.

Matched pair analysis

Cubbage (1983) investigated matched pairs selected so that for every owner who had been advised by a State Cooperative Forest Management Program (CFM) forester a matching property was chosen in the vicinity with similiar growing stock, area and other physical variables but no such contact. As the author of the study pointed out, it might be that the results instead of indicating the effect of promotion might reflect the impact of some difference in the CFM population that distinguished them from non-CFM ownerships. But this very proper caution applies *a fortiori* to the sample interviews where the selection of promotion presence or absence might have masked a difference which was not in fact related to the effect of extension itself.

Aggregate time series analysis

All the above studies used data for individual forest ownerships in a given period as the sample observations. Tikkanen (1981) and de Steiguer (1983) reported analyses regressing investment in reforestation on income and other variables plus planting grants. Unfortunately extension was not incorporated into the regressions which possessed few enough degrees of freedom in both cases, but the studies are mentioned in order to indicate that such aggregate approaches are feasible: they run the usual risks of misidentification of underlying factors influencing behaviour.

Conclusions

The foregoing references on the effectiveness of extension show a mixed set of results. Cubbage *et al.* (1993) note that in the light of such evaluations in the early 1980s more were conducted in other States but with mixed results. In some cases this was because of poor specification of the model, as in Washington where variation in log quality dominated any effect that might have derived from extension, and in Minnesota where extension is superfluous in any event in aspen stands since these have only to be cut and coppice allowed to regrow so rendering the impact of extension trivial. Inferences drawn have generally been in favour of real though only modest gains from promotion.

There are several reasons why workers in this field feel unable to make more than inferences about the impact of extension. In particular, the prospect is that those who choose to ask for technical advice *prima facie* provide evidence of a special interest. Educational level may be judged useful in controlling for this kind of effect but few studies have incorporated such a variable. Those with more knowledge of forestry matters were found by Riihinen (1970) to be less interested in cooperation, planning and recommended operations, and a broadly similar result emerges from the later work of Karppinen and Hänninen (1990); namely that the progressive owners were ones who were negative towards promotion.

Second, it is harder to identify the impact of extension than of, say, price or grant level because information, which is not an all-or-nothing input, is difficult to measure. Despite these observations, the evidence for useful gains from promotion is important, especially as in several cases the level of pay-off from this activity appeared to be higher than that from grant payment.

The Cutting Decision

An obvious area of interest to government which has assisted the creation of a growing stock, to industry long promised increased quantities of wood and to harvesters, is the willingness of owners to cut wood from their current holding of growing stock. Several policy measures may be deployed to influence such action. Taxation, advice, market information and direct subsidies for thinning operations and roading are the principal tools.

Some of the papers referred to above deal with the harvesting decision, the importance attached to this in the United States stemming from the fact that this term refers to clear felling and leads to the question of whether and if so how to restock.

Several Nordic studies have considered the cutting behaviour of owners. Partly arising out of the major effort devoted to this subject in Sweden, Carlén (1990) has been able to make a number of tests of the possible effects of factors. He concludes that, in all of the features tested, actions of forest owners in the sample of over 1000 for whom he had data seemed to support the conclusion that the behaviour of private owners was highly rational. Rudqvist and Tornqvist (1986) and, based on their data, Lönnstedt (1986) made a detailed study of 40 ownerships scattered throughout Sweden. They attempted to identify which of several factors accounted for the different forestry strategies adopted. These include the individual owner's tradition of farming and forestry, current employment and time available for forestry work, and the financial situation of the household. These are well-known features which have often been found, in combination with factors of the economic environment, to have strong influences on cutting decisions. An important feature is the cyclical course of private ownership, leading to different attitudes over time to logging by an individual owner. Lönnstedt calculated a supply function for the cut in terms of stumpage price and stumpage price change. As is well recognized, such relations are intensively behavioural in nature and the actual coefficients can be influenced by many factors. Lönnstedt provides a useful assessment of routes to encouraging increased cut in a given price and cost environment. These include alterations in fiscal and legal arrangements. Thus, changes in tax rules on allowable deductions and national insurance charges were believed likely to increase the cut. Of greater novelty, he suggests that forest bonds with a guaranteed rate of interest could be a useful means of introducing more flexibility by assurance of income stability.

A broadly similar discussion can be found in Kuuluvainen and Salo (1990) who assert that in a perfect capital market, owners would be expected to decide the cut in the light of interest rates, growth and present and future prices. Drawing attention to the variation in the cut with age of owner, from 4.1 m^3 ha^{-1} for owners under 35 years to 1.8 for those over 65, the authors note that with a perfect capital market the cut would be expected to be constant with age. That the cut does not actually vary as predicted suggests that there are imperfections in the capital market, forest owners using their woods as a bank even if they could borrow or lend through the banking system. This appears to be too strong a hypothesis, since, for example, the utility assigned by an owner to holding growing stock rather than a bank credit may change with age, but the analysis is most useful in drawing attention to aspects that can be expected to have a bearing on the rate of cut in circumstances where the income derived from forestry forms a significant part of total income.

Tikkanen and Vehkamäki (1990) develop a supply relation similar to that of Lönnstedt. This incorporates more variables including interest rate, the owner's net asset position and tax exemptions available. They deduced that forestry taxation and management planning might be expected to have little effect on cut. To the extent that small yet significant impacts on cut might be derived from changes in agricultural policy (so affecting income in this activity) and in taxation rules, it is interesting to find two Finnish forest economists stating that they believed that prospects for gaining agreement to such changes were low.

Surprisingly few studies have been undertaken on the impact of thinning subsidies such as those offered in a number of European countries. However, Baardsen (1991) has considered the whole range of grants and subsidies available in Norway from the point of view of the social benefit calculated as the discounted value of marginal volumes mobilized at current standing tree values. In the case of the subsidization of first thinning introduced from 1985, he finds the effectiveness fairly high. Carlén and Löfgren (1986) studied the short-term subsidization of thinning in northern Sweden which was undertaken at the behest of the pulping industry and which lasted for only one year. They concluded that it stood out as one of the most ineffective uses of government money for forestry in that country.

Other Applications

Finally, some studies of particular institutions may be noted. Hultkrantz (1987) considered the possible gains from deregulation of the rural land market in Sweden. The expectation of such action was that there would be an opening up leading to purchase by generally more active and enterprising owners. The conclusion was, however, that this would not increase the short-run supply of wood, which was assumed to remain a major goal of policy in Sweden. Also Lönnstedt's (1991) analysis of the impact of forest authority rules (such as minimum felling age and maximum allowable area under 20 years of age on any given property) on yield and hence revenue stands out as a rare example of enquiry in a field which deserves more effort.

Conclusions

Overall, it may be concluded that satisfactory tools exist for assessing the factors bearing on owners' decisions and thus deducing the extent to which different policy tools affect their choices. This means that the effectiveness of such policy measures can be evaluated. The assessment of the economic efficiency of schemes is more difficult. But ability to progress in the latter field depends partly on the fact that most programmes are intended to increase timber supply and the economic magnitudes relevant to estimation of effects on wood price are broadly known; only the discount rate remains troublesome.

Where the programme is concerned with environmental effects in addition to wood production and where knowledge of the various demand and price elasticities is poor, the chances of calculating social gains or losses of programmes are much reduced. Yet it is probable that technically determined rules will increasingly be promulgated for the supply of environmental benefits. The work quoted in this appendix depends on the availability of quantity and price information. The need for more quantification, which in turn calls for the development of suitable mensurational methods for the various outputs, and valuation of environmental benefits is clear.

References

Adams, D.M. and Haynes, R.W. (1980) The 1980 timber assessment market model: structure, projections and policy simulations. *Forest Science* 26(3), Monograph 22.

Arnold, J.E.M. (1991) The long term global demand for and supply of wood. Paper No 3, *Forestry Expansion: a Study of the Technical, Economic and Ecological Factors*. Forestry Commission, Edinburgh.

Baardsen, S. (1991) [Economic results in forestry – an effectiveness analysis.] *Skogforsk*, Nr. 7, Ås.

Binkley, C.S. (1981) Timber supply from private nonindustrial forests: a microeconomic analysis of landowner behaviour. *Yale School of Forestry and Environmental Studies Bulletin* no. 92, Newhaven.

Boyd, R. (1983) The effects of FIP and forestry assistance on nonindustrial private forests. In: Royer, J.P. and Risbrudt, C.D. (eds), *Nonindustrial Private Forests: A Review of Economic and Policy Studies*. Symposium Proceedings, Duke University School of Forestry and Environmental Studies, pp. 189–203.

Boyd, R.G., Daniels, B.J., Fallon, R. and Hyde, W.F. (1988) Measuring the effectiveness of public forestry assistance programs. *Forest Ecology and Management* 23(4), 297–309.

Carlén, O. (1990) Private nonindustrial forest owners' management behaviour. Swedish University of Agricultural Sciences, *Department of Forest Economics: Report* 92, Umeå.

Carlén, O. and Löfgren, K.-G. (1986) Supply consequences of subsidizing thinning activities in Sweden. *Scandinavian Journal of Forestry Research* 1, 379–386.

Clawson, M. (1979) *The Economics of Nonindustrial Private Forests*. Resources for the Future, Washington, DC.

Cubbage, F. (1983) Measuring the physical effects of technical advice from service foresters. In: Royer, J.P. and Risbrudt, C.D. (eds), *Nonindustrial Private*

Forests: A Review of Economic and Policy Studies. Symposium Proceedings, Duke University School of Forestry and Environmental Studies, pp. 252–258.

Cubbage, F., O'Laughlin, J. and Bullock, C.S. (1993) *Forest Resource Policy* (Chapter 17, Public assistance for private forest owners). Wiley, Chichester.

De Steiguer, J.E. (1983) The influence of incentive programs on nonindustrial private forestry investment. In: Royer, J.P. and Risbrudt, C.D. (eds), *Nonindustrial Private Forests: A Review of Economic and Policy Studies*. Symposium Proceedings, Duke University School of Forestry and Environmental Studies, pp. 157–164.

Gansner, D.A. and Herrick, O.W. (1980) Cooperative forestry assistance in the Northeast. *USDA Forest Service Research Paper* NE-464.

Gray, G.J. (1990) Evaluation of forest resource planning activities of forest resource activities of US State governments. *Proceedings, Division 4*, XIX IUFRO Congress, Montreal.

Haynes, R.W. (1990) An analysis of the timber situation in the United States: 1989–2040. *USDA Forest Service General Technical Report* RM-199.

Hultkrantz, L. (1987) Deregulation of the rural land market – a way to increase forest harvesting in Sweden? Swedish University of Agricultural Sciences, *Department of Forest Economics Report* 65, Umeå.

Hyde, W.F. and Newman, D.H. (1990) The value of price information in roundwood markets. *Proceedings Division 4*, 265–277. XIX IUFRO Congress, Montreal.

Järveläinen, V.-P. (ed.) (1981) Effectiveness of forest policy on small woodlands. Contributions of Working Party 4.06-01, IUFRO. *Silva Fennica* 15(1), 23–111.

Järveläinen, V.-P. (1986) Effects of forestry extension on the use of allowable cut in non-industrial private forests. In: Tikkanen, I. (ed.), Analysis and evaluation of public forest policies. *Silva Fennica* 20(4), 312–318.

Kallio, M., Dykstra, D.P. and Binkley, C.S. (1987) *The Global Forest Sector: an Analytical Perspective*. Wiley, New York.

Karppinen, H. and Hänninen, H. (1990) *An Information System for Monitoring Private Forestry*. Voluntary paper, S 4.08-01, XIX IUFRO Congress, Montreal.

Krygier, J.T.and Deneke, F.J. (1983) Cooperative Extension Service programs for the education of forest landowners. In: Royer, J.P. and Risbrudt, C.D. (eds), *Nonindustrial Private Forests: A Review of Economic and Policy Studies*. Symposium Proceedings, Duke University School of Forestry and Environmental Studies, pp. 259–267.

Kuuluvainen, J. and Salo, J. (1990) Nonindustrial private forest owners' timber supply and life cycle harvesting in Finland. *Proceedings Division 4*, 278–289. XIX IUFRO Congress, Montreal.

Lönnstedt, L. (1986) [Cutting decisions by non-industrial private forest owners – an analysis of case studies.] *Report* no 172, Department of Operational Efficiency, The Swedish University of Agricultural Sciences, Garpenberg.

Lönnstedt, L. (1991) Economic consequences of the Swedish forest policy. Department of Forest Industry Market Studies. Typescript. Uppsala.

Marty, R. (1983) Retargeting public forestry assistance programs in the North. In: Royer, J.P. and Risbrudt, C.D. (eds), *Nonindustrial Private Forests: A Review of Economic and Policy Studies*. Symposium Proceedings, Duke University School of Forestry and Environmental Studies, pp. 281–287.

Riihinen, P. (1970) The forest owner and his attitudes toward forestry promotion. *Acta Forestalia Fennica*, 109.

Risbrudt, C.D. and Kaiser, F. (1982) Economic analysis of the sawmill improvement program. *Forest Products Journal* 32(8), 25-28.

Royer, J.P. (1987) Determinants of reforestation behaviour among Southern landowners. *Forest Science* 33(3), 654-667.

Royer, J.P. and Risbrudt, C.D. (eds) (1983) *Nonindustrial Private Forests: a Review of Economic and Policy Studies*. Symposium Proceedings, Duke University School of Forestry and Environmental Studies.

Rudqvist, A. and Tornqvist, T. (1986) [Forestry strategies and cutting decisions among private forest owners - 40 case studies.] *Report* no. 171, Department of Operational Efficiency, The Swedish University of Agricultural Sciences, Garpenberg.

Solberg, B. (1986) Forest sector simulation models as methodological tools in forest policy analysis. *In: Royer, J.P. and Risbrudt, C.D. (eds), Nonindustrial Private Forests: A Review of Economic and Policy Studies*. Symposium Proceedings, Duke University School of Forestry and Environmental Studies, pp. 419-427.

Tikkanen, I. (1981) Effects of public forest policy in Finland: an econometric approach to empirical policy analysis. In: Järveläinen, V.-P. (ed.) Effectiveness of forest policy on small woodlands. *Silva Fennica* 15(1): 38-64.

Tikkanen, I. (ed.) (1986) Analysis and evaluation of public forest policies. Contributions to Working Party 4.06-01, IUFRO. *Silva Fennica* 20(4), 251-427.

Tikkanen, I. (1986) Search for innovative forest policies and programs the future and role of policy and program analysis. In: Tikkanen, I. (ed.), Analysis and evaluation of public forest policies. *Silva Fennica* 20(4), 253-427.

Tikkanen, I. and Vehkamäki, S. (1990) The effects of economic and forest policy on roundwood supply. *Proceedings Division 4*, 350-360. XIX IUFRO Congress, Montreal.

USDA Forest Service (1982) An analysis of the timber situation in the United States 1952-2030. *Forest Research Report* no. 23. Washington, DC.

Young, R.A. and Reichenbach, M.R. (1987) Factors influencing the timber harvest intentions of nonindustrial forest owners. *Forest Science* 33(2), 381-393.

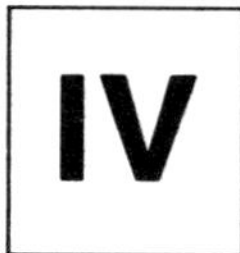

STERLING EXCHANGE RATES[a]

Country	Currency	Rate in January 1992	Rate in October 1992
Australia	A. dollar	2.3	2.25
Belgium	B. fr.	56.83	50.72
Denmark	D. kr.	10.67	9.53
France	Fr. fr.	9.35	8.35
Germany	DM.	2.75	2.45
Ireland	IR£	1.02	0.92
Netherlands	gld.	3.1	2.76
Norway	N. kr.	10.82	10.35
Sweden	S. kr.	10.07	10.03
Switzerland	Sw. fr.	2.44	2.19
United States	$	1.75	1.57
EC	ECU	1.4	1.24

[a] Rates in national currency units per £ sterling are shown for two months in order to reflect the variation which arose in 1992. Values are the averages of middle rates for each working day. Source: Midland Bank plc.

INDEX